Small Fry

Britain's Tiniest Freshwater Fishes

Small Fry
Britain's Tiniest Freshwater Fishes

Mark Everard

with photographs by **Jack Perks**

CABI

CABI is a trading name of CAB International

CABI	CABI
Nosworthy Way	200 Portland Street
Wallingford	Boston
Oxfordshire OX10 8DE	MA 02114
UK	USA
Tel: +44 (0)1491 832111	Tel: +1 (617)682-9015
E-mail: info@cabi.org	E-mail: cabi-nao@cabi.org
Website: www.cabi.org	

ISBN-13: 9781836991670 (hardback)
         9781836991687 (paperback)
         9781836991694 (ePDF)
         9781836991700 (ePub)

DOI: 10.1079/9781836991700.0000

Commissioning Editor: Jamie Lee
Editorial Assistant: Theresa Regueira
Production Editor: Shankari Wilford

Typeset by Straive, Pondicherry, India

Contents

Introduction: Small Wonders

Fishes are wildlife! Until very recently, fish have had a far lower profile among those with interests in wildlife when compared with other taxa. Freshwater fishes have been particularly overlooked or perhaps assumed as of primary interest only to anglers or as things for other wildlife or people to eat. This is true in many countries across the world, as it has been in Britain. Yet fish are not only diverse in appearance with fascinating life habits but also play integral roles in complex ecological networks with other organisms from the microbial to the largest of mammals (Fig. 1.1).

We humans also have an odd habit of ignoring the small things in life in favour of the large. Consequently, the smaller fish species found throughout our diverse freshwater bodies – farm ponds and other pools, lakes as well as gravel pits and reservoirs of all sizes, canals and drainage channels and ditches, flowing water ranging from mighty salmon rivers to often-neglected brooks and estuaries – suffer doubly from neglect.

Yet the 'small fry' species found in British fresh waters are not only hugely varied but exhibit intriguing life histories. They are also often quite beautiful. In addition, they play significant ecological roles connecting levels of food chains as both small predators but also as prey, and as agents in the regulation of parasites, pests and diseases. Small freshwater fishes are also creatures that are

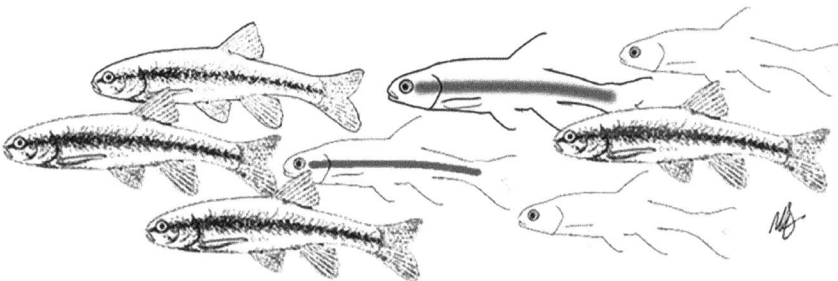

Fig. 1.1. A shoal of minnows. (Image © Mark Everard.)

© Mark Everard 2025. *Small Fry: Britain's Tiniest Freshwater Fishes* (M. Everard)
DOI: 10.1079/9781836991700.0001

particularly captivating to children, some of whom grow up to fall in love with and dedicate themselves to conserving or restoring planetary ecosystems. That is certainly my life story in a nutshell, or perhaps that should be in a puddle.

It is not that big fishes aren't wonderful, their gilded scales glinting under a bright sky and with fins held erect as they mouth air while you admire them a moment before slipping them back into the mysterious depths none the worse for the experience. These necessarily brief moments of contact in our air-filled surroundings with scarcer denizens of a wholly different watery domain add so much to one's sense of wonder and well-being, providing tangible and inspirational connections with the supportive and beneficial processes of other unseen worlds.

However, big fish are also rare beasts, at least in any natural water. Progressive mortality throughout life means that the billions of eggs that are fertilized produce millions of fry, from which thousands may survive the first summer and hundreds or less the first winter. Of these survivors, the toll of predation, disease and other hazards may deliver perhaps a few tens to survive long enough to shed their roe and milt, thereby to propagate the species. They grow large in only ones and twos, rare giants representing the capstone of the wide pyramid that is a population of fish.

And, among them, though often unheeded, swim a diversity of smaller species, busy about their fascinating life cycles and survival strategies that keep aquatic ecosystems functioning.

The Small Foundations of Mighty Pyramids

For all the charisma of megafauna, the great ecosystems of Earth are driven not by the massive but by the minute and the imperceptible. On East Africa's great plains, the enormous biomass of large mammals is awe-inspiring, though dwarfed by the cumulative bulk of tiny termites almost entirely unseen beneath their feet. These innumerable termites in turn are rendered inconsequential against the weight of microbes and other invisible creatures underpinning the dynamics of the grasslands. In fact, microbes are thought to make up half of total global biomass. This pyramid of numbers ranging from the biggest and most charismatic species right down to the microscopic is also observed in the world's oceans, where the aggregate weight of giant squid and immense whales is almost inconsequential compared with the combined biomass of tiny krill, up to two hundred million tonnes sometimes within a single shoal, and these krill in turn depend upon a planktonic soup comprising a vastly greater bulk of living matter that, though individually far tinier, cumulatively accounts for 70% of the biomass of the marine environment.

All fish species, both larger and smaller, are intimately interdependent players in aquatic ecosystems. The bulkier, ever-hungry omnivores and predators depend upon vastly greater numbers of smaller fish and other prey which, in turn, browse upon a countless resource of invertebrates, algae, plants and microbes. As with the population of any one fish species, the base of the whole ecosystem's pyramid is broad, with tiny and less visible species accounting for the overwhelmingly greatest overall biomass and also performing the bulk of nature's processing and cycling, supporting everything that is built upon it.

It is the inconsequential that captures and converts the overwhelming load of the energy and matter on planet Earth, maintaining it in endless bio-geochemical cycles underpinning nature's continuance. It is, ultimately, the vast throng of minuscule players that supports such necessarily scarcer and cumbersome organisms as human beings, big fish and indeed almost all visible living things. Fascinating though big fish and other of the planet's megafauna may be, a cold fact is that Mother Nature would not stop without them. And, humbling though it may be for our self-centred human psychology, the nat-ural world would cope perfectly well without us humans too. Indeed, given the mess that we have made of her, she may just be better off shedding her burden of bipedal parasites! The big fish, as for big species of all kinds, are inestimably less important than our own self-importance, and rest upon the foundation provided by the smallest and generally most neglected of the planet's life forms.

The little things, including little fishes along with a diverse assemblage of algae and other plants and invertebrates, are far more than mere pawns and prawns; the objective reality is that they are the biggest players in the game of keeping Earth's ecosystems running.

What is a Small Fish?

Fish fry are among the smallest of all vertebrate animals. They hatch in a semi-embryonic state, many around the size of an eyelash, still connected as larvae to their yolk sac and with fins, guts and bones largely undeveloped into their early free-swimming days. Some species grow large, and others less so. It is the 'small fry' species that interest us here.

What qualifies as a 'small fry' species? In truth, any division of 'small spe-cies' from larger fishes is arbitrary. In fisheries circles, there is a term 'minor species' that I find unhelpfully patronizing given the vital roles that all fish spe-cies play, especially those smaller players at the broad base of the pyramid as, though diminutive, they are all none the less of great importance for the func-tioning and resilience of ecosystems and as unique and wonderful as fishes of any other size. Better, I think, to express the arbitrary nature of this distinction as those species that do not exceed 1 pound (454 grams) at fully adult weight.

We knew intimately how fascinating the small fishes were as kids during those tender and untrammelled times when all and any fish or living thing could inspire pure joy. We were then swathed by a glorious innocence that held us in thrall to the stiff, zigzag dance of the nearly ubiquitous three-spined stickleback, male fish as gaudy and bold as popinjays. We marvelled at the curiosity of black-lined minnows as they thronged to nibble our toes in clear streams, scattering as we moved yet jostling back in an instant to nibble away once more. The Tommy ruffe was a familiar acquaintance to many of us, ever eager to snaffle a worm in all weathers, raising indignant, pointy fins as we hauled it wriggling from the water. Many of us will have waded urban or rural streams, turning over stones to trap bullheads and stone loaches in our chilled and scratched hands. Almost without exception, we'll have hunted the margins of pools and rivers with net and jam jar throughout the long, seemingly endless summer days of our youths when life and the times in which we lived were so much simpler. And, of course, gudgeon; those iridescent marvels! Who among us older, more cynically world-weary anglers are not still instantly excited as a sprightly 'gonk' comes to hand?

Reconnecting with Small, Everyday Wonders

When we were barely fry ourselves, senses yet to be dulled by familiarity, our eyes were open to the profusion of small wonders that swam in our local pools and streams. In those far-off days, we were alert and alive to the great power of the small and the rarity inherent in the commonplace.

As we age, we may rely more on revelations in the visual media or from science to remind ourselves about how fantastic these small worlds remain. Or perhaps we feel we need to validate our now-muted childish fascination through some spurious 'grown-up' rationality? All the while, whether they are seen, appreciated or not, these abundant small life forms remain busy keeping the world's ecosystems ticking over. Rediscovering our enjoyment of them can help keep us sane in an ever-busier world obsessed with the idolatry of gigantism. 'Bigger', 'faster', 'more'... where will it all end? Certainly not with a greater appreciation of simplicity, sufficiency and the things that are of greatest importance for keeping this tired old Earth turning, and we with it.

Once, the smaller fishes were a source of far more than sport alone. They also provided nourishment, with many small species appreciated on the plates of rich and poor alike. Meticulously organized gudgeon-fishing parties were a highpoint of the social calendar in Victorian England. Some of these little fishes had additional economic importance, with sticklebacks harvested to fertilize fields in eastern England, and the scales of the humble and often-maligned bleak were once exploited for the lucrative manufacture of artificial pearls.

A healthy obsession with little fishes opens our eyes to other residents of water bodies, including their rich fishy, mammalian, bird, plant and other wildlife. Suddenly, a tiny forest stream, overgrown pool, disused canal or unprepossessing ditch draws you in to explore their potential for hosting little fishes prospering from neglect and lack of predation from their bigger cousins, in addition to their wealth of amphibian, insect, plant and other life. To the inquisitive, no body of water, however apparently inconsequential, is too small to rouse curiosity about the fishy and other living wonders that it may hold (Fig. 1.2). Eyes and attention may also wander inshore from the deep and open

Fig. 1.2. No body of water, however apparently inconsequential, is too small to rouse curiosity. (Image © Mark Everard.)

water where bigger fish may reside, enquiring of the shallow and weedy margins about the small wonders they may conceal.

Small life forms – slime, bugs, beasts and small fry – quietly but diligently remain ever-busy, dedicated to their singular yet interdependent roles that, cumulatively, keep the whole world spinning. Ironically, refocusing on a world in microcosm opens the mind to nature's greater vistas.

Celebrating Britain's Wealth of Small Freshwater Fishes

Few books have been dedicated to this world of smaller but fascinating freshwater fish species. One of the most memorable for me was Kenneth Mansfield's delightful *Small Fry and Bait Fish: How to Catch Them*, published in 1958: coincidentally the year in which I was born and a major influence on me as a small fry myself.

Then, we had to wait another half-century before publication in 2008 of my own *The Little Book of Little Fishes*, now sadly long out of print. I have also written two more books dedicated solely to little fishes, both republished in 2023 – *Gudgeon: The Angler's Favourite Tiddler* and *Ruffe: The Spiky Freshwater Ruffian* – in addition to ensuring that these tiddlers are included in various of my other more general books on fish. I have also extolled their virtues in television shows and radio programmes. My hope is that this new book throws more overdue limelight on these small and often-neglected freshwater wonders.

There is science sprinkled judiciously here and there across these pages, and the odd Latin name too, but only just sufficient for the diligent to follow or the disinclined to ignore.

Whatever your predilection, we are all certainly well overdue reacquainting ourselves with the little fishes that make life rich, possible and fun (Fig. 1.3).

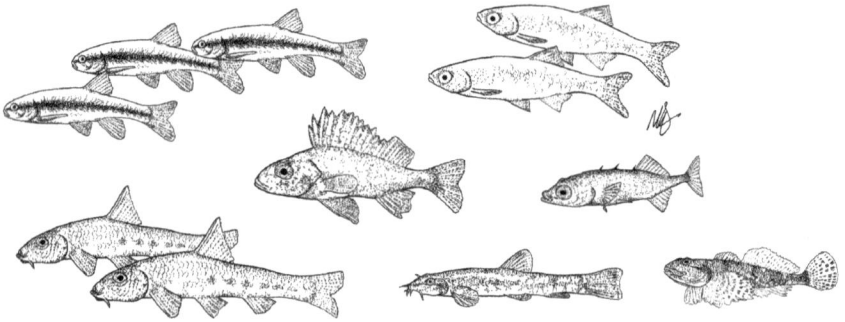

Fig. 1.3. A smorgasbord of little fishes! (Image © Mark Everard.)

Minnows: The Archetypal 'Tiddler'   **2**

The minnow (Fig. 2.1), common and widespread in Britain and across much of Europe, is the epitome of the term 'tiddler'. Found in rivers and small streams, many canals and larger lakes, the minnow is the classic quarry of the 'net and jam jar' hunter that many of us were as mere small fry ourselves.

They are also irrepressibly curious. Wiggle your toes in a summer stream and, chances are, a shoal of minnows will come to investigate. The sensation of minnows nibbling your toes and leg hairs is truly invigorating. Move suddenly and they will skitter away, but not for long as their curiosity drives them back once again to explore this strange presence in their watery domain. While I am pretty certain that I would be far less emboldened were I a mere bite-sized snack for many aquatic predators, you can almost imagine – as you can readily also observe among heifers put out to graze in the spring – shoal members egging each other on to explore this strange new manifestation in their world!

It is true that, for the angler, minnows can mob a soft bait intended for larger quarry, stripping it back to a bare hook in next to no time. But hold on a minute and spare a kindly thought for our sprightly little black-lined friends. Bless them all: where would we be without them?

Minnows made our younger days, and the days of younger people still, full of wonder. This was not just in the capture of these ever-willing feeders but, through them, to unveil and view afresh an alien world of magic and mystery.

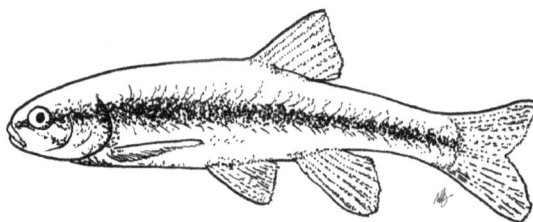

Fig. 2.1. The minnow. (Image © Mark Everard.)

© Mark Everard 2025. *Small Fry: Britain's Tiniest Freshwater Fishes* (M. Everard)
DOI: 10.1079/9781836991700.0002

Most of us will recall, and relive through kids today, the first simple joy of a float made to skitter under the attentions of unseen but clearly ravenous fishes, no matter what their size, with minnows often a first encounter. Minnows may have ignited the fires of our early angling and wildlife interests and fed the flames increasingly consuming us into our mature years. Catching or simply seeing a fish of any size matters in those embryonic days, and the ever-reliable minnow often fills that need admirably!

Without the presence of minnows in many of our waters, what would the prospects be for the vitality of the wider ecosystem or for catching the bigger fish on to which budding anglers progress? After all, minnows are among the most ubiquitous links in food chains connecting microscopic and tiny plants and invertebrates to the larger food items upon which larger species depend. Minnows too, along with many 'small fry' fish species, form a significant part of the diets of kingfishers, little and great-crested grebes, egrets, gulls, dippers and goosanders that are part of the vista that makes time spent by the water so enriching. They feed predatory fish too, as well as piscivorous mammals, without which the waterside would be a far poorer place. It is instructive but also inspiring to pause and consider the meticulous mechanisms of nature played out by all their constituents, from the charismatic to the humble and often overlooked. All constitute vital tiers in river and still-water ecosystems, upon which the whole complex fishery depends.

Quite aside from their piscatorial and ecological importance, the humble minnow has inspired stories – mentioned in the works of Beatrix Potter among others – that have meaning for people, even if most folks would not be able to confidently identify a minnow if one fell out of the sky into their lap! Minnows are also the subject of recipes, having fed poor people or supplemented the diets of many more in years gone by.

There are also political dimensions, as a fishless water just looks 'wrong' to the lay person, and public disquiet about the widespread abuse of our rivers can form part of an upwelling of activism drawing political attention to the need to better protect our precious natural inheritance. These unsung heroes thereby contribute to the quality of life of us all.

So, if you are an angler – I guess many of the readers of this book may be – then please spare a kindly thought for these small workers when next you are visited by a plague of minnows snaffling baits set out for larger fishes. Better still, take a young friend along with you to see the world, and these archetypal 'tiddlers', through eyes less jaded by the cynicism of the adult world!

Natural History of the Minnow

What is a minnow?

This may seem a deceptively simple question, but the term 'minnow' is applied to many species right across the world from Australia to China. What we are talking about here, in a British Isles context, is the European minnow, which goes by the Latin name *Phoxinus phoxinus* (Linnaeus, 1758). This is the full Latin name, denoting that this species was classified by the 'father of taxonomy' Carl von Linné (Linnaeus in Latin) back in 1758. But the simple name 'minnow' will suffice in this chapter once we have cleared up exactly which fish we are talking about. This minnow is in the minnow or dace-like fish family (Leuciscidae). (A brief history of the relatively recent distinction of the minnows and dace as a distinct family is outlined in Box 2.1.)

European minnows are shoaling fishes, reaching a maximum length of some 14 centimetres (about 5½ inches) though they are generally far smaller, with a maximum reported age of 11 years. The body is rounded in cross-section and streamlined, generally of a torpedo-like shape, coloured brown on the back and silver or white beneath for at least most of the year. The dark line along the flanks of the streamlined, rounded body will be a familiar characteristic to many, though this 'line' in fact comprises a row of vertically elongated blotches that commonly fuse into each other. While appearing scaleless, the bodies of minnows are actually evenly covered by very fine cycloid scales.

The fins are short, rounded and colourless. For those interested in technical details, the dorsal fin is held erect by three spines followed by six to eight soft branched rays (denoted III/6-8), the anal fin is III/6-8 and the caudal (or tail) fin is 0/19. The mouth is small and terminal, neither upward- nor downward-facing, and no barbels (sensory 'whiskers') are present. As for all cypriniforms (fishes in the order Cypriniformes), minnows also lack teeth in the mouth,

---

Box 2.1. Taxonomic details of the minnow with a note on the splitting of the family Leuciscidae

---

Dipping lightly into some more scientific terminology, minnows are part of the class of ray-finned fishes (Actinopterygii). Within this class, they are classified in the order of carp-like fishes (Cypriniformes).

For many years, some 3160 species of fishes in 376 genera were lumped into a broad and long-established 'minnow and carp' family (Cyprinidae). This former minnow and carp family comprised often quite dissimilar fishes of widely varying forms inhabiting fresh waters (only two species occur in fully marine waters) from North America, Africa and Eurasia. However, so diverse was this family of minnows and carps that, based on morphometric (physical characteristics) and genetic evidence, the family was split into a range of different families. Four of these families occur in Britain.

Among these subdivided families is the Leuciscidae (the true minnows or dace-like fishes), within which the Eurasian minnow fits along with a range of other species encountered in Britain including roach (*Rutilus rutilus*), rudd (*Scarndinius erythrophthalmus*), chub (*Squalius cephalus*), dace (*Leuciscus leuciscus*), common bream (*Abramis brama*), silver bream (*Blicca bjoerkna*) and bleak (*Alburnus alburnus*).

Within the family Leuciscidae, there are six subfamilies including the Phoxininae (the 'true minnows') typified by the European minnow.

though food is crushed by pharyngeal teeth in the throat that evolved as modifi-cations of the gill arches.

Throughout the spring and summer, minnows can form dense aggre-gations and also often betray their presence by leaping from the water. Close observations reveal that this leaping behaviour is not generally to escape pred-ators and is most evident into evening.

Minnow distribution, habits, diet and senses

Minnows are widely distributed across Eurasia, including the drainage basins of the Atlantic Ocean, the North Sea, the Baltic Sea, the Arctic Ocean and freshwater catchments of the northern Pacific Ocean (including the Amur drainage basin and Korea).

They can occur throughout Great Britain but are locally absent in some less accessible regions. The natural range of the minnow, as for many British fish species, results from historic connections between British rivers and con-tinental catchments. These river basins were once connected by a land bridge known as 'Doggerland'. Towards the end of the ice age, between 6500 and 6200 BC, a gigantic ice lake to the north was breached causing a megatsunami that inundated the land bridge, which now constitutes the Dogger Bank be-neath the southern North Sea. Prior to the inundation of Doggerland, the rivers Thames, Rhine and Scheldt converged to form the Channel River – today the English Channel – carrying their combined flow to the Atlantic. Impenetrable watersheds formerly prevented the northward and westward spread of fresh-water fishes lacking marine life stages. Many such fish species were naturally limited to drainage basins from what is now the Humber in the north to the Thames in the south. However, a wide range of species, including minnows, have subsequently been widely distributed by human agency across Britain, both accidentally and deliberately.

Minnows were not naturally present in the island of Ireland as, lacking a historic land and river bridge connection, minnows were unable to colonize. Fishes historically present in Ireland were documented by Giraldus Cambrensis ('Gerald of Wales', *c*.1146–*c*.1223), a medieval clergyman and chronicler of his times, in his book *The History and Topography of Ireland*. Giraldus Cambrensis recorded that "pike, perch, roach, gardon, gudgeon, minnow, loach, bull-heads and verones" were absent from Ireland. Of the freshwater fishes found in Ireland recorded by Giraldus – brown trout (*Salmo trutta*), Atlantic salmon (*Salmo salar*) and arctic charr (*Salvelinus alpinus*) as well as pollan (*Coregonus lavaretus*), three-spined sticklebacks (*Gasterosteus aculatus*), European eels (*Anguilla anguilla*), smelt (*Osmerus eperlanus*), shad (*Alosa* spp.), three species of lamprey (*Petromyzon marinus*, *Lampetra fluviatilis* and *Lampetra planeri*) and the (far from common) common sturgeon (*Acipenser sturio*) – all are tolerant of salt water either as migratory or brackish-water species. However, minnows now occur locally in Ireland, along with many other non-native species, resulting from deliberate or unintentional introductions from mainland Britain.

Minnows are generally fishes of flowing or well-aerated waters, including in wave-lapped margins of lakes and pools. They are, though, uncommon in completely stagnant systems. Minnows are also omnivorous, feeding on a range of small animals and plants as well as diffuse organic matter (detritus).

In common with many fish species, minnows have well-developed sight and also chemical senses that are registered not merely though nostrils but with sensors spread across the body enabling them to respond to gradients of taste/smell (these senses blur in aquatic environments) in the water column.

One of the more prominent senses in minnows, as indeed many other fishes, is the detection of pressure waves in the water via the lateral line system. The lateral line comprises a series of sensory organs in a linear arrangement along the flanks, detecting vibrations and pressure gradients in surrounding water. The lateral line of the minnow runs continuously from the rear of the gill cover to the caudal peduncle (the 'wrist' immediately in front of the tail fin). Within each of the sensory pits comprising the lateral line are hair cells responding to alterations in pressure, discerning low-frequency sounds, surface waves and currents. The lateral line system enables fish to navigate their environment, even in darkness. Experimentally blinded minnows are still able to navigate by sensing currents around objects in their environment. The lateral line also enables fish to detect their predators as well as playing roles in shoaling behaviour, mating and aggression. The lateral line of the minnow is complete, running along the 80–100 small cycloid scales (thin, rounded scales arranged in an overlapping pattern) along the mid-line of each flank of the torpedo-shaped body from behind the operculum (gill cover) to the tail.

Of particular interest is that minnows also detect the 'smell' of their predators. Some predatory fishes release pheromones (hormones released eternally to the body) and it is these that minnows have been found able to detect, modifying their behaviour accordingly. The term 'kairomones' is used to describe pheromones to which other species react. Predatory pike (*Esox lucius*) have been found to adapt their behaviour by defecating in areas remote from those in which they feed, separating the more concentrated release of kairomones from the areas in which they hunt and where it is preferable for them for their prey species to be less vigilant.

Another pheromonal response that has been observed in minnows, as well as some other fishes, is an adverse reaction to 'schreckstoff' (literally 'fright substance' in German as the term was coined by an Austrian scientist). Schreckstoff, also known as 'alarm substance', is released when the skin of a fish is damaged and 'alarm substance cells' (or ASCs) are broken. The scent of schreckstoff triggers a fright reaction in other fishes, which change their physiology and behaviour – hiding, shoaling more tightly, reducing activity, ceasing to feed, heightening awareness of danger and physiological stress reactions – to reduce their vulnerability to predation. Minnows exposed to schreckstoff at a specific location in experimental aquaria tend to avoid feeding at that spot, and also to avoid it for several days. This demonstrates learning behaviour.

Minnow reproduction and development

Minnows spawn in shoals on the marginal gravels of rivers and large lakes, and can do so multiple times throughout the late spring and summer. At these times, males and females often form discrete shoals that come together to spawn. It is in spawning condition that male minnows become spectacularly gaudy, perhaps the most ornamental of all native freshwater fishes, developing white patches at their fin bases and with a kaleidoscope of emerald, red and gold colours across the body. By contrast, females retain their overall silver-brown coloration throughout the year. Once eggs are released over flushed gravel, males shed their milt to fertilize them. There is no parental care.

Hatchlings initially have a yolk sac attached to their bellies but, once this is consumed, they emerge as larval fish with as-yet incompletely developed fins and gut. Reliant on tiny invertebrate and algal food items and shelter from strong flows, they undergo several stages of metamorphosis. Their diet as they grow to adult size comprises opportunistic feeding on a wide range of small invertebrates, plants and detritus matter.

Minnow as prey and predator

Many fish, bird and mammal species prey opportunistically on minnows. Kingfishers are well known as predators of 'small fry' species, with minnows and also the fry of larger fish species constituting common food items. Egrets as well as cormorants, herons, mergansers and goosanders and other piscivorous birds also make use of their seasonal abundance. Otters may not actively hunt these small and agile fishes but can consume them opportunistically. Many fishes, including such predators as pike, perch (*Perca fluviatilis*), zander (*Sander lucioperca*) and brown trout, readily consume minnows, as do opportunistic species such as chub and barbel (*Barbus barbus*).

Despite their small stature, minnows can be voracious predators too. This is particularly so when raiding the spawn of other fishes of running water and lake margins where present. When chub and barbel are observed spawning on flushed gravels, these larger fishes are often attended by shoals of minnows ready to consume eggs washed down in the current or exposed in the gravels into which females deposit their spawn. Though this may appear as carnage, in the grand scheme of spawning with no parental care, the losses are made up by sheer numbers of eggs laid.

Minnows and Angling

Few people other than young children set out deliberately to fish for minnows as sport. That said, they can be great fun on light tackle, and I am one of the few that do enjoy a minnow session from time to time! Small floats, light lines and hooks of size 20 or smaller are ideal, with maggots a mouth-sized bait, though, in reality, any bait will suffice including snippets of bread and small red worms. No specialist tackle is required; line tied to a hazel switch 'rod' cut from the hedge adds to the rustic charm of a minnow session!

Minnows can also be collected as bait for other fish species, both known predators and opportunists such as chub and barbel. They can be caught in numbers on fine rod and line, by net, or – my favourite method and one from my childhood – with a baited minnow trap (Fig. 2.2). To make a classic minnow trap, take a wine bottle with a pronounced punt (the dome in the bottom) and carefully tap out the top of the punt using a hammer and thick nail. Put some bread in the trap, insert the cork and tie string around the bottle to retrieve it. Then, lay it on a gravel bed where minnows are present with the corked neck facing upstream. Soon, the ever-curious minnows will explore this new item and try to get in to eat the bread. Minnows approaching from the downstream end swim up into the dome, entering the bottle. Seeking corners and walls to escape, they then find themselves trapped. The actions of these minnows attract more minnows and, before you know it, the trap is full of minnows! But do not leave the trap *in situ* for long as otherwise these small fishes may become overcrowded and deplete the oxygen in the water. Other types of traps can be made from large jars with some form of inverted cone in the top, be that a funnel or the neck of a plastic bottle with the access hole enlarged.

Minnows and Society

Other names by which minnows are known

Minnows also go by other common British names. One such is the 'penk', also the 'common minnow' and the 'Eurasian minnow'.

Fig. 2.2. A classic wine bottle minnow trap with the bottom cone punched out. (Image © Mark Everard.)

They also go by a wide variety of names in the languages of people inhabiting their wider geographical range, a subset of which include:

French	Vairon, Charbonnie, Gendarme, Petit blanc, Sprille
Dutch	Elrits
German	Balte, Bambel, Bambeli, Bammeli, Bampel, Bitterziimpelchen, Blitterfischl, Blutelritze, Brunnenpfrill, Budd, Burli, Buthe, Butt, Ellering, Ellritzeritze, Elritze, Gefrille, Haberfischl, Pfrille, Piek, Rümpchen, Spierling, Weiling, Zaukerl
Danish	Almindelig elritse, Elrits, Elritse, Europæisk Elritse
Swedish	Äling, Elritsa, Kvidd

I have no explanation as to why the minnow has quite so many alternative names in German, those in the list above being just a subset of a far longer list of common names!

'Minnow' is also commonly applied as a term for the diminutive, be that a small person or a sports team "like minnows" facing a bigger and better-resourced adversary.

Minnows and the arts

Minnows have swum into the works of some notable artists. One such whose work has become familiar to many is the Irish-born British watercolour artist Alexander Francis Lydon (1836/1837–1917), who produced classic artworks of British freshwater fishes. These famously illustrate the Reverend W. Houghton's 1879 book *British Fresh-Water Fishes* and have been widely reproduced and used for many purposes, including for example by the Portmeirion Pottery company for their *Compleat Angler* range of tableware. Lydon's painting of the minnow is part of a collage of a number of other fishes (Fig. 2.3).

Another set of images familiar to many older anglers and wildlife enthusiasts will be those produced as tea cards by the Brooke Bond company in 1960 (Fig. 2.4), the artist for which was not disclosed by the company. None the less, these images are widely appreciated and still frequently reused.

Fig. 2.3. Painting of a minnow by the artist A.F. Lydon.

Fig. 2.4. Painting of a minnow appearing in a set of 1960 Brooke Bond tea cards.

The minnow also appears in some stories. One of the more popular and well-known of these is *The Tale of Mr. Jeremy Fisher*, written by Beatrix Potter and published in 1906. Jeremy Fisher was a frog who set out to catch a minnow for his dinner,

> "I will get some worms and go fishing and catch a dish of minnows for my dinner", said Mr. Jeremy Fisher. "If I catch more than five fish, I will invite my friends Mr. Alderman Ptolemy Tortoise and Sir Isaac Newton. The Alderman, however, eats salad".

After sitting down to eat a butterfly sandwich until a shower passed, Jeremy Fisher had an unintended bycatch of little Jack Sharp, the stickleback, but ended up nearly eaten by a trout. (Spoiler alert: he was saved only by his mackintosh!)

Minnows as food

Just one example of a classic culinary use of this common freshwater species is the 'minnow tansy', as recorded in Izaak Walton's classic 1653 book *The Compleat Angler*:

> And in the spring they make of them excellent Minnow-tansies; for being washed well in salt, and their heads and tails cut off, and their guts taken out ... being fried with yolk of eggs, the flowers of cowslips and of primroses, and a little tansy; thus used they make a dainty dish of meat.

Some other former angling authors have extolled the virtues, when they were children, of pickling minnows whole for later consumption. In his 1873 book *Familiar History of British Fishes*, Frank Buckland wrote that,

> I used to pickle minnows in vinegar and spice, and keep them in pickle bottles. They were capital eating, especially when, as a 'junior', I was not over-fed and had to look out for what extra 'grub' I could get hold of.

Some others recommend that cleaned minnows, with no need to remove the bones, can be fried as whitebait. Kenneth Mansfield was not entirely euphoric about their gastronomic virtues though, writing in his 1958 book *Small Fry and Bait Fish: How to Catch Them* that,

Minnows have never been popular as food in Britain but they were (and still probably are) cooked and served as whitebait in some families. Most boys have tasted them in their youth, but the primitive methods of cooking were not enough (in my case, at any rate) to impart a lifelong taste for minnows.

Minnows and nature conservation

The IUCN Red List (The International Union for Conservation of Nature's Red List of Threatened Species), documenting the extinction risk of different species globally, classifies minnows as of 'Least Concern' (LC) reflecting their common and widespread presence. The widespread distribution of minnows does not warrant their explicit inclusion on any of the schedules of the various strands of UK and European nature conservation legislation.

The value of minnows in the diets of a wide range of other wildlife, including many species of birds, mammals, fish and other groups, means that the importance of their presence should not be overlooked. Minnows are crucial to the overall vitality of rivers; as I have said on television before, "Lose the minnows and we lose the lifeblood of the river" (Fig. 2.5).

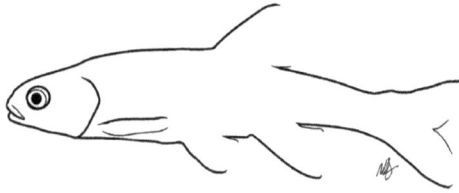

Fig. 2.5. Outline of a minnow. (Image © Mark Everard.)

Gudgeon: The Angler's Favourite Tiddler

3

The title of this chapter – 'Gudgeon: The Angler's Favourite Tiddler' – mirrors the title of my book on the gudgeon, the only book dedicated to this charismatic and rather wonderful species (Fig. 3.1). But the title also reflects the fact that so many anglers simply love catching gudgeon. Perhaps they do not set out to do so deliberately, though some of us do, but as a welcome visitor to a day's sport. I do not know why, but gudgeon always make me smile, and this sentiment is shared by so many other fish-lovers.

Gudgeon are, in truth, rather splendid fish, among the most beautiful that swim British and European waters. Their flanks, covered by a chain mail of large and conspicuous rounded scales, may initially appear brownish. But hold them in the light and there are spots of different muted hues on the background silvery sheen interspersed with darker patches, all overlain by a subtle violet iridescence. Quite marvellous to behold! There is something appealing too about the large eyes set astride a short, rounded head. The small underslung mouth is adorned at each corner by a single fine barbel. The streamlined body shape, generally round in section but flattened beneath reflecting life on the bed, is creamy beneath and darkening up the flanks to the back.

Gudgeon are, in fact, so common and widespread, from fast rivers to smaller streams, canals, commercial fisheries and ponds, that they are readily overlooked. This is particularly so in murky, overstocked water where they are

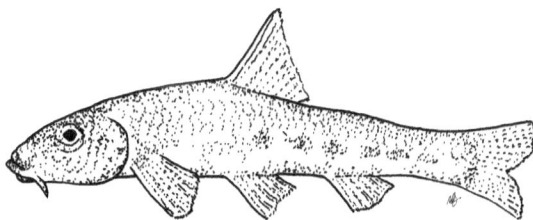

Fig. 3.1. The gudgeon. (Image © Mark Everard.)

© Mark Everard 2025. *Small Fry: Britain's Tiniest Freshwater Fishes* (M. Everard)
DOI: 10.1079/9781836991700.0003

often of insipid colour and small size. But they remain fishes for which many of us 'grown-up' anglers retain a childlike, possibly irrational love. Just watching a shoal of shoal of gudgeon working over a shingle run beneath the rod top, their heads down and grazing together like so many sheep, brings a smile. They can be relied upon to feed throughout the coldest of days, when they are particularly welcome as other more torpid fishes neglect your bait.

My first gudgeon were caught from Kent's River Medway back in the early 1960s in a stretch that was a childhood paradise of wildlife and wonder but has long since been bypassed and left to silt up in the name of 'progress'. This is a source of great sadness, perhaps also for those long-lost, innocent days. Fishing the same river again 60 years later, the capture of two gudgeon in a mixed bag was joy unalloyed! Gudgeon have this power to connect with the child within us all. Is it that they were common when we were young, or is it their cartoon-like rounded face and two large eyes? Whatever the reality, these hungry little jewels welcome us back time and again to the riverbank, and we love them for it!

Gudgeon do not grow huge, nor are they rare. Despite that, they are almost universally liked, even loved, by anglers of all ages and those interested in the fishes of British fresh waters. Quite why they are so widely loved is a mystery ... they just are!

Natural History of the Gudgeon

The gudgeon will be familiar to many British readers of this book. However, the common name 'gudgeon' has been applied to a wide variety of fish species across the world. This includes the unrelated topmouth gudgeon (*Pseudorasbora parva*), an invasive alien species in British waters that is covered later in this book. Some 40–50 species of freshwater fishes distributed across Australia are also commonly known as 'gudgeon', including some species growing up to 3 kilograms (6.6 pounds) though most are considerably smaller. However, all of these Australian 'gudgeon' species are in the goby family (Gobiidae), quite unrelated to the freshwater gudgeon familiar to us in British waters.

What is a gudgeon?

The gudgeon considered here is the species going by the Latin name *Gobio gobio* (Linnaeus, 1758). Gudgeon fall taxonomically into the order of carp-like fishes (Cypriniformes), part of the gudgeon family (Gobionidae). There are 216 species in the Gobionidae family, all of them inhabiting fresh or brackish waters, but only one, *G. gobio*, occurs in British waters.

Gudgeon can reach a maximum length of some 21 centimetres (about 8¼ inches), though most fish encountered are far smaller than this. The maximum reported weight is 220 grams (nearly 8 ounces), though British specimens reach substantially smaller sizes. The largest gudgeon are females. The maximum reported age for a gudgeon is 8 years, females reputed to live longer than males.

Gudgeon have an elongated body shape, tapering from a head that is approximately round in cross-section towards a laterally compressed tail. The body is flattened beneath, reflecting the benthic (living on the bed) habit of this fish. The flanks are covered by a chain mail of large and conspicuous scales, with a silvery background colour that is darker greeny-brown on the back fading to creamy-white beneath, mottled with spots and patches of varying muted hues, all overlain with a delicate violet iridescence. Body colour varies enormously with habitat, with fish from murky canals often appearing silvery with little more elaborate coloration. There are between 38 and 45 scales along each flank, the scales on the mid-line perforated by a complete lateral line.

The gudgeon's head is short, with prominent eyes and a small mouth that is downward-oriented, reflecting its habit of feeding on the bed of rivers and pools. In common with other cypriniforms, gudgeon have no teeth in the mouth but possess pharyngeal teeth in the throat to crush food items before they pass down into the gut. There is a single barbel (or 'whisker') at each corner of the mouth to help gudgeon probe for small food items, a feature that is also useful for differentiating gudgeon from young barbel (*Barbus barbus*) that have two pairs of barbels. The barbel is also a more robust fish, with a proportionately smaller eye and it is also often darker in colour (Fig. 3.2).

Fig. 3.2. Comparative photographs of the heads and mouths of gudgeon (above) and barbel (below). (Image © Mark Everard.)

The number of barbels around the gudgeon's mouth can also distinguish very small gudgeon from the two British loach species – the stone loach (*Barbatula barbatula*) and the spined loach (*Cobitis taenia*) discussed later in this book – each of which possesses three pairs of barbels.

The fins of the gudgeon are generally short-based. The dorsal (back) fin has a spine/ray formula of II-III/5-7 (two or three spines followed by five to seven soft branched rays) and, like the caudal (tail) fin and the pectoral fins, is generally heavily spotted. The anal fin has a formula of II-III/6-8 and is generally less coloured, as are the ventral fins, both on the gudgeon's underside.

Gudgeon distribution, habits, diet and senses

Gudgeon are widely distributed across many northern European drainage basins, though absent from Norway and the north of Sweden and Finland. They also occur eastwards across northern Asia as far as the Korean Peninsula. Though naturally absent from Italy, Ireland (their former absence from Ireland is for the reason described for the minnow), Wales and Scotland, they have none the less been locally introduced and become naturalized. Equally, though naturally absent from Greece and the Iberian Peninsula, gudgeon have become established locally in these places following introduction.

Like many species of fish, gudgeon were naturally present only in the river catchments of eastern England historically connected to continental European drainage basins via the 'Doggerland Bridge' up to the end of the last ice age (between 6500 and 6200 BC). However, gudgeon have since become more widely distributed across mainland Britain due to translocations by people, both deliberately and as passengers when other fishes have been relocated, including widely across England except for the far south-west. They are also now locally distributed across Scotland and Wales.

Gudgeon prefer, and generally prosper best in, rapidly running water, and have a pronounced preference for sandy or gravel beds. However, they can also proliferate in slow-flowing lowland streams, canals and many still waters. During the warmer months of the year, gudgeon tend to gather in small shoals in shallow water where they are found in the greatest densities, whereas in winter the fish retire to deeper waters. Gudgeon also sometimes occur in brackish water but are not tolerant of higher salinities.

The benthic habit and body form of gudgeon lead them to feed primarily on the bed of the water bodies in which they occur. They can also feed on vertical and steep banks, such as those found in canals and reservoirs, as well as on submerged surfaces such as sunken woody material. They are omnivorous fish, feeding on a range of small animals and plants as well as diffuse organic matter (detritus). Favoured foods comprise small invertebrates, including various worms and insect larvae such as bloodworms (the bright red larvae of chironomid midges often found in muddy beds) as well as crustaceans and molluscs, and also, occasionally, fish eggs and even fish fry.

As also described for the minnow, the eyesight, lateral-line pressure sensors and chemical sense of the gudgeon are well-developed. Less is known about their responses to pheromones.

Another fascinating feature of the gudgeon is that these small fishes are also vocal. Gudgeon can emit squeaking vocalizations, believed to be for communication with other gudgeon. Vocalizations vary with the degree of activity of the fish, which is itself proportional with temperature. However, as vocalizations occur independent of the spawning season, they are thought not to have a significant role in breeding. It is considered likely that they support shoaling behaviour or that they enable gudgeon to alert others of their kind to danger, though these speculations are unproven.

Gudgeon are also sometimes in the habit of leaping clear of the water in warmer conditions, their bodies vibrating rapidly with an audible buzz that can be heard on a quiet evening. The reason for this is not clear – it could be an escape reaction, to rid some external parasites or simply to burn off excessive energy – but this behaviour has been noted by a range of angling authors.

Gudgeon reproduction and development

Gudgeon typically spawn once per year in many waters, generally between mid-April and July, though in highly productive water bodies they are known to spawn multiple times. Sandy beds are favoured as a spawning substrate. Female gudgeon release between 1000 and 3000 eggs at intervals over several days above harder substrates. The eggs drift in the current and are fertilized by milt released in synchrony by male gudgeon. At this time, male gudgeon can develop a dense covering of horny spawning tubercles on their heads and the front of the body. Gudgeon eggs measure 2–5 millimetres (0.08 to 0.2 inches) in diameter, often appearing to have a bluish tinge. The eggs are sticky, adhering on contact with the bed of the stream or pool. A. Lawrence Wells describes the fate and the appearance of gudgeon eggs after release in his 1941 book *The Observer's Book of Freshwater Fishes of the British Isles*,

> Sinking to the bottom and there adhering to the stones, so that the flow of water will not wash them away, they may be seen as the sun glints across them like tiny soap bubbles.

Conversely, Adrian Pinder, in his 2001 book *Keys to Larval and Juvenile Stages of Coarse Fishes from Fresh Waters in the British Isles*, reports that he has never found gudgeon eggs on sand (though acknowledging that they would be hard to spot there) but that he collected them from mossy growths on well-flushed surfaces.

After the eggs are released, gudgeon exhibit no parental protection. On hatching, typically after 10–30 days depending on water temperature, juvenile gudgeon are incompletely developed and are still attached to a yolk sac. Metamorphosis continues after the larvae become free-swimming. Juvenile gudgeon remain on the bottom, preferring habitats of low current speed and a detritus-rich sandy bed providing a wealth of small food items to support their continuing development. Juvenile gudgeon tend to shoal on the bed of the water body near the places where they were spawned.

Growth rate varies substantially between different types of water, responding to variable factors including temperature and the availability of food. Growth is rapid during the first two years of life, juvenile gudgeon reaching lengths of up to 5 to 6 centimetres (up to 2 or more inches) over this time period, though growth slows subsequently. Both males and females attain a length of around 10 centimetres (4 inches) in their third year on maturity, though some females may mature in their second year. Gudgeon are not long-lived fishes: fish of 6 and 7 years old are uncommon, and the maximum recorded age is 8 years.

Gudgeon as prey

Gudgeon, as many small fishes, have a wide range of predators. They are consumed by numerous fish species that are principally predatory, including perch, pike, zander, chub, trout and eels. However, various other omnivorous fish species, such as barbel, opportunistically feed on gudgeon small enough to engulf when they are encountered, and particularly so in warmer weather when their metabolism is at its most rapid.

Gudgeon also readily fall prey to various species of piscivorous bird. Kingfishers predate small fry of many species, pouncing from a high vantage point into the water to grasp individual fish before returning to the perch to stun and devour them. Cormorants, herons and egrets too can and do hunt gudgeon and other small fishes, as do saw-billed ducks (mergansers and goosanders).

Otters also consume gudgeon and other small fishes opportunistically. Otters typically favour bottom-dwelling species, also including bullheads, loach and eels, that they can trap with their stone-turning feeding habit, gudgeon falling less easily as prey as they are more associated with open beds.

Gudgeon and Angling

Kenneth Mansfield was a great champion of the little fishes, and his delightful 1958 book *Small Fry and Bait Fish: How to Catch Them* was a major inspiration to me when I was but a small fry myself. Of the gudgeon, Mansfield wrote that:

> Gudgeon are the most important fish mentioned in this book from the angler's point of view, for they provide sport, make very good live baits, serve the match fisherman well and are extremely good eating. Their nearest rival in these respects is the bleak.

Economically, gudgeon are particularly important for match fishing, a staple catch on some waters. This fish is often also prized as a target by pleasure anglers. In his 1926 book *Fresh-water Fishing for the Beginner*, Arthur P. Bell described the gudgeon succinctly and affectionately as,

> A SPORTING little river fish.

Location and attraction of gudgeon

Gudgeon are strongly attracted to disturbance of the riverbed. This may be due to natural conditions such as the activities of animals, but they also respond to interventions by people. Mr Lane, in his 1843 book *The Diary of A. J. Lane: With a Description of those Fishes to be found in British Fresh Waters*, advised the gudgeon angler to,

> Go to shallows and gravelly places free from weeds or where cattle have gone in to drink.

Many older angling writers throughout centuries have advocated raking river-beds to attract gudgeon and to stimulate them to feed. This advice dates back centuries, including, for example, in the 1613 book *The Secrets of Angling* by John Dennys (believed to be a fishing companion of William Shakespeare) that records:

> In the Thames they always use a raik [sic] (and by raiking the ground and discoloring the water occasionally) draw them together.

Similarly, in his 1943 book *Coarse Fish*, Eric Marshall-Hardy wrote that,

> If the angler will take the trouble to rake the bottom regularly while fishing they may be caught by the hundred...

To this advice, Kenneth Mansfield wrote in this 1958 book *Small Fry and Bait Fish: How to Catch Them* that:

> Once a shoal of gudgeon has been located, by observation or by trial-and-error angling, rake the bottom.

An alternative to raking the bed is, of course, putting in ground bait to add some cloud and flavour to your swim, optionally mixing fragments of worm or maggots to condition these fishes to seek out the bait on your hook. Disturbing the stream bed with your feet, dislodging find food particles and a silt trail, can also draw in these curious fish.

Gudgeon baits

Understanding the gudgeon's diet is obviously important for selecting the best bait to put on the hook. As we have seen, gudgeon are opportunist feeders on the beds of rivers, streams, lakes, ponds and canals, particularly in shallower regions at least during warmer months. Gudgeon also have a marked preference for sandy bottoms; when I have scuba-dived in rivers, I have observed that when I find a sandy patch, often quite localized, that is where I will also find gudgeon.

Their catholic tastes mean that gudgeon can be caught on a range of small baits, both animal and vegetable. As insect larvae are an important element of the gudgeon's diet, various varieties of maggots are particularly effective baits. Some match anglers favour dead maggots over live ones, possibly as they do not rapidly bury themselves in the sediment on the bed of the water. Bloodworms, the larvae of midges, are also accepted by gudgeon though this is probably an unnecessarily expensive option. Small worms, or fragments of larger worms, are also eagerly taken, noting the tendency of gudgeon to eat 'meaty' baits. As long ago as 1496, the English nun and writer Dame Juliana Berners wrote in her *Treatyse of Fysshynge with an Angle* that [my interpretations in square brackets],

> The Gogen is a good fysshe of the mochenes [for its size]; and he biteth wel at the grounde. And his baytes for all the yere ben thyse: ye red worme: codworme [caddis]: and maggdes [maggots].

Likewise, Mr Lane wrote in his 1843 book *The Diary of A.J. Lane: With a Description of those Fishes to be found in British Fresh Waters*,

> Bait red worms or gentles. [Author's note: As many older anglers will remember, 'gentles' was a common name for maggots at least through to the 1960s.]

On a similar theme, Henry Coxon wrote in his 1896 book *A Modern Treatise on Practical Coarse Fish Angling* that,

> They are fond of gravelly bottoms and are easily caught, either with a bit of a worm, or a gentle.

Small fragments of prawns can also be effective, as can small cubes of luncheon meat particularly in warmer conditions when gudgeon seem to have a stronger preference for meaty baits. However, so too can bread presented as a small piece of flake, punched discs or paste on a fine hook. In reality, a hungry gudgeon is unlikely to refuse any type of bait small enough to engulf.

However, bear in mind always that gudgeon are bottom-dwelling fishes. Putting the bait on the bed where they will find it is probably far more important than the type of bait on the hook.

Gudgeon tackle and tactics

When fishing for gudgeon, there is no specialist angling advice, or particular tackle requirement, beyond using a small hook under light float or leger tackle suited to the waters in which you are fishing, making sure that the bait is on or very close to the bed. Further important advice is to enjoy every capture of this charismatic little fish! As Kenneth Mansfield put it in his 1958 book *Small Fry and Bait Fish: How to Catch Them*,

It is unnecessary to go into long details of tackle and methods.

The most important aspect is that the bait has to be presented hard on the bottom. A bait any significant distance above the bed will be ignored by gudgeon, no matter how many of them are present. This fact is behind one of the endearing local common names of this fish – 'plummets' – as the angler can keep adjusting the depth at which their bait is presented below a float until they catch a gudgeon, confirming that they have reached the riverbed!

The ease with which gudgeon may be caught on a range of tactics led Izaak Walton, in his 1653 book *The Compleat Angler*, to note that,

He is an excellent fish to enter a young angler ...

Advice on gudgeon tactics and baits holds true even for 'specimen-sized' gudgeon – the British record at the time of writing is a magnificent fish of 142 grams (5 ounces) taken in 1990 by D.H. Hull from the River Nadder (a tributary of the Hampshire Avon) at Sutton Mandeville, Wiltshire.

Gudgeon and match angling

Unlike many, but not all, of the 'small fry' species covered in this book, gudgeon can be important to the match angler. Where present in numbers, and where larger species are less abundant, gudgeon can become a prime target for aggregating a weight of fish, potentially with a few 'bonus fish' such as roach, bream, dace or chub also locating the angler's bait.

Speed is of the essence to the competent match angler, intent on drawing gudgeon in closer through targeted feeding. The bait is then presented on whip, pole, running line float or light leger tactics on a short line so that hooked fish to be swung rapidly to hand. Gudgeon may be particularly important in more challenging weather and water conditions when other fish are less inclined to feed, but when gudgeon seem less perturbed by these otherwise adverse conditions.

Gudgeon as bait

As gudgeon are eagerly taken by predators of all kinds, they have also been widely used as baits for predatory fish species. This is also due to the facts that they are hardy and abundant locally. As Eric Marshall-Hardy wrote of gudgeon in his 1943 book *Coarse Fish*,

... not to speak of their usefulness as live baits for Pike, Perch and Chub (the smallest for the latter).

Perch and zander anglers in particular favour gudgeon presented as a live bait. Eels and opportunists such as barbel will also readily accept a gudgeon, presented dead or alive. For live-baiting, the unfortunate gudgeon (for which I have sufficient empathy not to do this!) is lip-hooked either with a large single hook or with a light snap tackle of paired treble hooks, and ideally allowed to rove freely under a float, a free-running leger rig or on a free line.

Izaak Walton also wrote in *The Compleat Angler* that gudgeon, as well as bleak and roach, were an ideal bait for pike, which could be attracted by the attentions of smaller fish if crumbs of white bread were sprinkled around the baited line. H.R. Robertson also wrote in his 1875 book *Life on the Upper Thames*, written during the heyday of Victorian England, that,

> Gudgeon are much used as bait when trolling for jack [small pike], and as a live bait for various large fish.

Gudgeon in the affection of anglers

From personal experience and drawing upon many things written about gudgeon, it is clear that very many people share a deep affection for this un-assuming little fish. I could not put it better than the words of Kenneth Mansfield in his wonderful little 1958 book *Small Fry and Bait Fishes: How to Catch Them*,

> One of the many surprising things in the sport of angling is the seemingly disproportionate delight that anglers of all ages in all ages have taken in catching gudgeon; fish whose average weight is about 3 oz.

There are in fact, at the time of writing, a number of gudgeon societies sharing the joy of this little fish. These include The Gudgeon Society, set up as a Facebook (social media) page by Carl Smith initially as a joke but at-tracting thousands of members. I had the honour of designing The Gudgeon Society's logo (Fig. 3.3).

To this jolly band of gudgeon-lovers we can add the Grand Union Gobio Gobio Society (GUGGS), formed in 2009 with the wonderful motto "Size doesn't matter ..." and dedicated to the pursuit and capture of canal gudgeon (initially from the Grand Union Canal but now extended to cover catching gudgeon from canals more generically).

Another like-minded gang of people with a passion for gudgeon is to be found among the Traditional Fisherman's Forum, comprising approximately

Fig. 3.3. The Gudgeon Society logo. (Image © Mark Everard.)

500 enthusiasts from the UK but also as far afield as continental Europe, America, Canada and South Africa.

The 'Jolly on the Wally' has been a regular event of the Traditional Angler's Forum since 2016, held on the tiny but delightful and mainly tree-covered River Wallington in south-east Hampshire, kindly hosted by the Portsmouth and District Angling Society, with anglers in the autumn match competing for the Gudgeon Jim Trophy in memory of the passing of Forum member 'Gudgeon Jim'.

Gudgeon and Society

Gudgeon are truly fascinating from many points of view – from ecology and wildlife-watching, in terms of their behaviour and for their angling interests – imbuing them with a range of meanings and cultural associations.

For anglers of all ages, the cheery appearance of a gudgeon, eager to feed in all weathers, almost unceasingly raises a smile. For the match angler, a shoal of gudgeon can bolster the net. For the 'fish twitcher' or even the casual nature-watcher, the sight of a shoal of gudgeon, together with their heads down grazing like so many sheep, is a joy to behold. Gudgeon are also widespread, inveigling their way into various dimensions of cultural life.

Other names by which gudgeon are known

The common English name 'gudgeon', first recorded in the fifteenth century, derives from Middle English 'gojoun', itself derived from the Middle French 'goujon'. The Latin word *Gobio* also means 'gudgeon' and is likely to be the origin of the Middle French term.

The word 'goujon', of course, is used in culinary circles to describe strips of fish, chicken breast or other meat that are coated in breadcrumbs and deep-fried. The shape of the fish is thought to have inspired the culinary term.

Other affectionate English nicknames for this charismatic little fish are 'gonks' as well as 'gobies' and, in the English Midlands, 'pongo'. When the Trent was a far more polluted river, gudgeon were also known as 'Trent barbel', though the river has since recovered and is now a top fishery with a good head of specimen barbel. The barbel-like comparison still persists, the gudgeon chapter of my 2008 book *The Little Book of Little Fishes* carrying the subtitle 'The poor man's barbel'. Affectionately, many people refer to gudgeon as 'gobies'. Also, as noted when considering angling for gudgeon, another amusing local name is 'plummets' as, if you keep adjusting the depth at which your bait is set below your float, you may eventually catch a gudgeon confirming you have reached the riverbed.

Across their range, gudgeon go by a range of local names, including:

French	Goujon, Gofi, Gouvin, Gouvion, Gressling, Grougnon, Grundling, Kress, Touret, Tragan, Troga
Dutch	Grondel, Riviergrondel
German	Giefen, Grässling, Gresling, Gressling, Grimpe, Gringel, Grüme, Grundel, Gründel, Gründling, Güwchen, Krasse, Webers
Danish	Almindelig grundling, Grundling
Swedish	Sandkrypare

Other aspects of gudgeon in language

Perhaps due to the widespread nature of the gudgeon and the non-exacting approach required to catch them, this fish has attracted a perception of simplicity that has led to the use of the term 'gudgeon' in that mildly insulting context. Take, for example, what was written by Eric Marshall-Hardy in his 1943 book *Coarse Fish*,

> the verb 'to gudgeon' crept into the English language in the 18th century, meaning roughly, 'to make a fool of by deception'.

However, use of the term 'gudgeon' denoting foolishness predates this by some centuries. No lesser a writer than William Shakespeare, who was known to fish with the angling author John Dennys, applied the foolish gudgeon as a metaphor in the *c.*1596 play *The Merchant of Venice*. The merchant of the play's title was Antonio, who famously made a contract with moneylender Shylock under the terms of which one pound of his flesh should be exacted if the loan was not repaid within three months. One of Shylock's many vocal and insulting critics was the character Gratiano who, during the trial, said to Antonio,

> I'll tell thee more of this another time.
> But fish not with this melancholy bait
> For this fool gudgeon, this opinion.

Samuel Butler, the seventeenth-century poet and satirist, in his poem *Hudibras* mocking Puritanism and the Parliamentarian cause from a Royalist perspective, wrote,

> Make fools believe in their foreseeing
> Of things before they are in being
> To swallow gudgeon ere th' are catch'd;
> And count their chickens ere th' are hatch'd.

A rather more baffling (in fact I have no idea what he is actually talking about!) recent reference is made by the Russian author Fyodor Dostoyevsky in his philosophical novel *The Brothers Karamazov*, published in instalments from 1879 to 1880,

> You save your souls here, eating cabbage, and think you are the righteous. You eat a gudgeon a day, and you think you bribe God with gudgeon.

Discussing the potential origins of the phrase 'to gudgeon a man' to denote the ease by which a person could be deceived, the Reverend W. Houghton considers in his 1879 book *British Fresh-Water Fishes* that this may have stemmed from an analogy by the English poet and dramatist John Gay (1685–1732), most fondly remembered for *The Beggar's Opera*, perhaps reflecting a misfortune of his own when writing,

> What gudgeons are we men,
> Every woman's easy prey!
> Though we've felt the hook, again
> We bite and they betray.

Gudgeon and the arts

Try as I have, I could not find any great opera, classical music composition or worthy work of literature dedicated wholly to the gudgeon – unlike, for example, the manner that the trout inspired Franz Schubert's lied (song) and quintet *Die Forelle* or how the Atlantic salmon stimulated Henry Williamson to write the novel *Salar the Salmon*.

However, there are many illustrations of gudgeon. These include my own drawings, some in this book as well as in my 2023 *Gudgeon: The Angler's Favourite Tiddler*, but also the works of many other illustrators as woodcuts, drawings and paintings throughout the centuries. Some out-of-copyright drawings feature in this book, and there are many more in publications on fish and fishing.

Of particular note are the paintings by A.F. Lydon, famously illustrating the Reverend W. Houghton's 1879 book *British Fresh-Water Fishes* and since widely reproduced for different purposes. Lydon's painting of the gudgeon is part of a collage with a barbel (Fig. 3.4).

Another set of images familiar to many older anglers will be those produced as tea cards by the Brooke Bond company in 1960 (Fig. 3.5), the artist

Fig. 3.4. Painting of a gudgeon by the artist A.F. Lydon.

Fig. 3.5. Painting of a gudgeon appearing in a set of 1960 Brooke Bond tea cards.

Fig. 3.6. A wonderful painting of a gudgeon by David Miller (left), which adorned the Environment Agency's 2018–2019 two-rod fishing licence (right). (Image © David Miller.)

for which was not disclosed by the company. None the less, these images are widely appreciated and reused.

An outstanding mention should go to the wonderful painting of a gudgeon by the fish and wildlife artist David Miller, which was used on the 2018–2019 Environment Agency two-rod coarse angling licence (Fig. 3.6). This and other of the myriad artworks dedicated to the gudgeon demonstrate the broad and enduring appeal of this small fish to the angling public.

The gudgeon has also featured on stamps in other European countries including, for example, in Sweden in 1991 and the closely related Danubian gudgeon (*Romanogobio uranoscopus*) on a Ukrainian stamp in 2019.

Gudgeon as pets

Gudgeon are one of the species of British freshwater fish that acclimatize quite readily to indoor aquaria, being hardy, not unduly large and unexacting in their water-quality requirements. They are a fascinating fish to watch, though you do have to keep the lighting subdued as they are easily spooked in bright or poorly planted tanks. To calm the gudgeon, it is worth putting them in with less-nervous species, for example crucians or sticklebacks, so that they feel more secure in venturing out from cover. I have found that the younger and smaller the gudgeon, the more rapidly they acclimatize to captivity, often investigating and feeding freely within the first day of introduction to their new home.

Given their predilection to seeking out stray scraps of edible matter of all sorts from the bed of the river or pool, Kenneth Mansfield wrote in his 1958 book *Small Fry and Bait Fishes: How to Catch Them* that,

> Some aquarium keepers like to have a gudgeon or two in their tanks to keep the bottom clean.

Gudgeon can from time to time be found for sale in garden centres for the home aquarist or pond-keeper, though at exorbitant prices. Gudgeon can be caught from the wild for this purpose, but you must of course ensure that you have the landowner's permission before you do this.

Gudgeon as food

Gudgeon have been widely praised for their culinary worth throughout history. From the days of the English Civil War, Izaak Walton wrote in his 1653 *The Compleat Angler* that,

> The GUDGEON is reputed a fish of excellent taste, and to be very wholesome.

Richard Franck noted in his 1694 book *Northern Memoirs* that,

> As the gudgeon is a most delicious fish, so ought he to be most delicately drest.

Mr and Mrs S.C. Hall also describe in their 1859 *The Book of the Thames from Its Rise to Its Fall* that,

> If people care to eat, as well as catch, fish, there is no fish of the Thames more 'palatable' than the gudgeon, fried with a plentiful supply of lard. It is 'of excellent taste, and very wholesome', and has been sometimes called 'the fresh-water smelt'.

Henry Coxon's 1896 book *A Modern Treatise on Practical Coarse Fish Angling* includes the comment that,

> Gudgeon are not only prolific fish, but are excellent for the table.

The Reverend W. Houghton was also effusive about the virtues of gudgeon in his 1879 *British Fresh-Water Fishes*, writing,

> In point of flavour the Gudgeon approaches that of the Smelt or Sparling, and in my opinion is one of the best of fresh-water fish we possess.

Arthur P. Bell was later to write in his 1926 book *Fresh-water Fishing for the Beginner* that,

> A dish of gudgeon fried a la whitebait is well worth eating.

Eric Marshall-Hardy was even more complimentary in his 1943 book *Coarse Fish*, stating,

There are, quite frankly, few coarse fish species that appeal to my palate, but
Gudgeon are very definitely an exception. I would rather by far a fry of these
delicious fish than a trout or grayling, and I have much experience of them all.

Angling author 'BB' (Denys J. Watkins-Pitchford) weighed in within an essay
'Food for Men', published in his classic 1946 *The Fisherman's Bedside Book* that,

The best eating of all coarse fish are gudgeon and perch.

Strong accolades indeed throughout history with, as we will see, a fry-up of
gudgeon one of the highpoints of Victorian society.

In times of need, gudgeon were also exploited when other food was scarce in
Britain: for example, throughout the Middle Ages and also during and following
World War Two when many coarse fish were still often taken to feed the family.

Gudgeon flesh is said to be finely textured as well as tasty, with recommendations that eight gudgeon per person suffices. Gudgeon are prepared by
first cutting off the head, cleaning out the guts and then thoroughly washing
and drying. The gudgeon are then rolled in flour, seasoned to taste, and fried
in butter or oil in a pan or deep-fried. This sounds delicious, but I am too soft-
hearted to want to inflict this on the gudgeon that I catch!

Gudgeon tansy was a popular recipe from the Middle Ages onwards.
Cleaned gudgeon were cooked with the bitter waterside herb tansy (*Tanacetum
vulgare*), similarly to the minnow tansy described by Izaak Walton. For those
wanting to delve deeper into recipes and recommendations for gudgeon, please
seek out 'BB's' 1987 book *Fisherman's Folly*, or Kenneth Mansfield's 1958
Small Fry and Bait Fishes: How to Catch Them, Eric Marshall-Hardy's 1943 book
Coarse Fish, or my own 2023 book *Gudgeon: The Angler's Favourite Tiddler*.

In France, gudgeon are highly prized for their culinary worth.

Gudgeon for the entertainment of English high society

Strange as it may now appear, fishing for gudgeon was a popular social pursuit
even by the upper classes in Victorian England. Supporting these high-society
expeditions, people of lower class were employed beforehand to rake the riverbed, creating a cloud of silt and exposing invertebrate food to attract these
inquisitive fishes. Ghillies (angling guides) would manoeuvre punts carrying
parties of angling guests, staking the vessels in position using ryepecks (stakes
in the riverbed), and then baiting hooks, usually with red worms, attached to
canes and fine lines. The duties of the ghillie also included unhooking their
fish (Fig. 3.7). Large catches could be made, and the fish were generally retained and subsequently cooked fresh as a feast for the parties of ladies and
gentlemen.

Although I have quoted this text from Kenneth Mansfield's 1958 book
Small Fry and Bait Fish: How to Catch Them in my 2023 book *Gudgeon: The
Angler's Favourite Tiddler*, Mansfield's recording of this social form of gudgeon
fishing warrants repetition,

Fig. 3.7. Woodcut of a Victorian-era gudgeon-fishing party.

> "in the 19th century it became a fashionable pastime on the Thames. Many people who never fished for anything else organised or took part in gudgeon-fishing expeditions on that river, hiring a punt and boatmen for the purpose. Amply supplied with food and drink, and with the professional Thames fisherman seeing to such practical matters as propelling the punt, adjusting the ryepecks, raking the bottom and baiting the hooks, the anglers, male and female, enjoyed in comfort the sport of gudgeon fishing.

> "Anglers of wider experience did not despise such expeditions, and many of the really famous fishermen of the day – Francis, Foster and Buckland among them – described the pleasure they gained from such convivial outings.

> "This picnic version of gudgeon fishing lost its popularity soon after the turn of the century and disappeared with many other pleasing idiocyncracies [*sic*] of the nineties. Real anglers pursued larger quarry and the little gudgeon was left to the young, the match fisherman, and the seeker after live bait."

The popularity of gudgeon-angling parties and the quantity of gudgeon caught are celebrated during the time when it was at its heyday in the chapter 'Gudgeon Fishing' in H.R. Robertson's 1875 book *Life on the Upper Thames*,

> Old anglers tell us that the gudgeon are on the decline in the Thames, both as to number and size. They 'remember the time' when eighty dozen were to be taken in the day by the party in one punt. Now, at the present time, in a take of fifteen or sixteen dozen, it is seldom a really sizable fish gets in the wells. If the extremity of the bye-laws of the fishery were carried out, every gudgeon fisher, as he carries away his fish, would be indictable for taking unsizable fish.

It has also emerged that there was some sense of these gudgeon-fishing parties being somewhat risqué, as ladies and gentlemen were often indulging in this piscatorial pastime without their usual chaperones. There was likely to have been rather more recreational interest in these less closely monitored expeditions than angling and dining alone!

Gudgeon and nature conservation

On the IUCN Red List (The International Union for Conservation of Nature's Red List of Threatened Species), documenting the extinction risk of different species, gudgeon are classified as of 'Least Concern' (LC) reflecting their common and widespread presence. Gudgeon are not specifically listed as of conservation concern in other European or British legislation, though the European Union (EU) Bern Convention (on the Conservation of European Wildlife and Natural Habitats 1979) does impose bans on a range of destructive fishing methods.

None the less, in common with other smaller fishes, gudgeon (Fig. 3.8) form important links in food chains constituting key elements of the biodiversity of freshwater habitats. This is important as freshwater ecosystems are regarded as among the most threatened ecosystems globally, with 38% of Europe's freshwater fish species threatened with extinction and a further 12 European fish species already declared extinct.

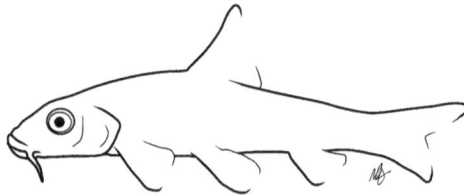

Fig. 3.8. Outline of a gudgeon. (Image © Mark Everard.)

Bleak: The Pocket-Sized 'Game' Fish

4

The bleak may be small (Fig. 4.1) but otherwise has all the attributes of a world-class game fish!

These sleek, shoaling denizens of the surface layers of rivers have an upward-oriented mouth, ever eager to engulf insects and other edible matter in the spindrift. Streamlined and exquisitely silvery, they share the profile of the much sought-after tarpon of tropical seas and can be readily caught on fly and bait alike.

Natural History of the Bleak

What is a bleak?

The bleak is a small fish, reaching a maximum length of 25 centimetres (approaching 10 inches) though is generally far smaller. This fish goes by the Latin name *Alburnus alburnus* (Linnaeus, 1758) and is part of the minnow family (Leuciscidae) along with such British freshwater fish species as roach, rudd, chub, dace, common bream, silver bream and, of course, the European minnow.

Bleak are shoaling fishes of flowing water. They have a seasonal habit of apparently disappearing during the winter – at this time they are far less active, forming aggregations in backwaters and other still reaches of water – then

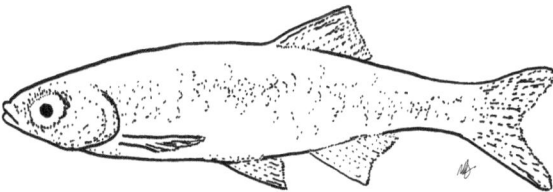

Fig. 4.1. The bleak. (Image © Mark Everard.)

© Mark Everard 2025. *Small Fry: Britain's Tiniest Freshwater Fishes* (M. Everard)
DOI: 10.1079/9781836991700.0004

reappearing in large numbers in the spring and summer, during which time they can form dense shoals. This seasonal reappearance lends them a common nickname of the 'river swallow'.

Bleak have sleek, laterally compressed bodies with brightly silvery flanks covered in prominent scales. The back is generally darker, commonly with a pronounced greenish tinge, fading through to white or cream on the underside. The mouth is large, lacking barbels, and is superior (upward-oriented). As with all other cypriniforms, bleak have no teeth in the mouth but instead have pharyngeal teeth in the throat to break down food items before they pass to the gut. This mouth orientation, together with the large and also upwardly situated eyes, is suited to the species' habit of feeding at the surface and in the upper layers of running water.

The fins of the bleak are colourless. The dorsal fin has the formula II-IV/7-9 (between two and four spines followed by seven to nine soft branched rays), and the caudal (tail) fin is supported by 19 soft rays. The base of the anal fin is particularly long with the formula III/14-20.

Bleak distribution, habits, diet and senses

Bleak are widespread across Europe north of the Caucasus, Pyrenees and Alps, and also across Asia eastwards to the Urals and Emba. Though naturally absent from the Iberian Peninsula and Italy, bleak have been locally introduced in Spain, Portugal and Italy. They are also naturally absent from Norway and Scandinavia north of a latitude of 67°N. Bleak are widespread in British rivers.

Bleak feed by intercepting fine food items in mid-water or at the surface, aided by their large upward-oriented mouths and eyes. Small aquatic and winged invertebrates form a significant proportion of the diet, but bleak will readily exploit small animal food items in the plankton as well as plant matter predominantly in the upper layers of rivers.

In addition to their acute vision, bleak have a complete lateral line running the length of each side of their flanks enabling them to respond minutely to river currents and vibrations, including those of their prey, shoal-mates and predators.

Bleak reproduction and development

Bleak spawn communally in large shoals during the spring on stones, gravel or nearby vegetation in shallow water. The sticky eggs receive no parental care once released. After hatching, bleak larvae remain in the margins of rivers and some of the larger wind-lapped lakes where they occur, moving out into open water to feed on plankton as they grow and mature. Juveniles feed extensively on small invertebrates and algae, whereas older fish feed opportunistically on small food items, particularly invertebrates, borne on the current or the water's surface.

Given their genetic closeness with other members of the Leuciscidae, hybrids are possible though rare due to different spawning-habitat requirements or timings across the seasons. Though scarce, hybrids between bleak and chub (*Squalius cephalus*) have been reported from British rivers.

Bleak and Angling

Bleak are not of specific interest to the pleasure river angler other than as part of a mixed day's sport and may sometimes be seen as a bait-robbing nuisance when present in dense populations during the warmer months. However, to the match angler, they can be a target for the rapid capture of many specimens to build a big weight.

Angling for bleak generally occurs up in the water using float tactics, be that presented on whip, pole or running line tactics. Regular loose-feeding with small baits such as maggots as well as bread or hemp can result in dense shoals of bleak competing for free offerings and losing any caution about baits presented on a small hook. For the match angler, speed is of the essence, so tight feeding of the shoal can draw them in close for rapid capture, unhooking and re-presentation of the bait under a light float.

For the more adventurous, bleak rise eagerly to a small dry fly, biting boldly and rapidly and presenting a great deal of fun on light fly-fishing tackle. This practice is ancient, Izaak Walton writing in his famous 1653 book *The Compleat Angler*,

> Or this fish may be caught with a fine small artificial fly, which is to be of a very sad brown colour, and very small, and the hook answerable. There is no better sport than whipping for Bleaks in a boat, or on a bank, in the swift water, in a summer's evening, with a hazel top about five or six foot long, and a line twice the length of the rod.

They will also accept small soft plastic lures presented on light and fine jig-head hooks, a relatively recent specialist method of fishing that is a gaining in popularity and for which you can find a great deal of useful details in the excellent 2019 book *Hooked on Lure Fishing* by Dominic Garnett and Andy Mytton.

Bleak, by virtue of their brilliant silvery coloration and size, are also an excellent bait for carnivorous fishes, albeit that they are not robust. Izaak Walton also wrote in *The Compleat Angler*,

> Take a small Bleak, or Roach, or Gudgeon, and bait it; and set it, alive, among your rods, two feet deep from the cork, with a little red worm on the point of the hook: then take a few crumbs of white bread, or some of the ground-bait, and sprinkle it gently amongst your rods. If Mr. Pike be there, then the little fish will skip out of the water at his appearance, but the live-set bait is sure to be taken.

Bleak and Society

Other names by which bleak are known

The Latin name *Alburnus* is said to derive from the city of Al Bura.

Strangely for such a common and widely distributed species, there are few if any common names for this freshwater fish across Britain other than the 'river swallow'. Izaak Walton also notes in his 1653 book *The Compleat Angler* the far from inappropriate name 'fresh-water sprat'.

Elsewhere across Europe, the common names applied to bleak are varied, including:

French	Ablette, Blanchet, Bleue, Coureur, Garlesco, Laube, Mirandelle, Nablé, Nablo, Ravanesco, Sardine
Dutch	Alver
German	Agoher, Agon, Agöne, Agune, Albala, Alve, Bestaller, Blacke, Bläke, Bläullg, Blieke, Blieken, Blinke, Butzli, Donaulauben, Gase, Grasle, Gris, Günger, Ikelei, Läge, Lang-Bleck, Langbleck, Laube, Lauben, Laubener, Lauber, Lauel, Lauge, Laugel, Laugele, Lauing, Laukel, Leiken, Maiblecken, Nestling, Ockelei, Okel, Pliete, Schaulaugeln, Schneider, Schneiderlein, Seelauben, Seeschiedl, Silberfisch, Silberling, Sonnenfisch, Sostknecht, Spitzlaube, Uekelei, Ukele, Ukelei, Uklei, Wiek, Wietling, Windlauben, Winger, Zumpel
Danish	Almindelig løje, Løje, Milling
Swedish	Benlöja, Löja

Once again, the German language excels with the huge range of local names, and variants of spelling, that it imparts on its small fish fauna!

Bleak and the arts

A.F. Lydon painted the bleak too in a collage with other larger fishes (Fig. 4.2), illustrating the Reverend W. Houghton's 1879 book *British Fresh-Water Fishes*, with his artwork subsequently widely reproduced and used for many purposes.

Another set of images familiar to many older anglers will be those produced as tea cards by the Brooke Bond company in 1960 (Fig. 4.3), the artist for which was not disclosed by the company. None the less, these images are also widely appreciated and reused.

Fig. 4.2. Painting of a bleak by the artist A.F. Lydon.

Fig. 4.3. Painting of a bleak appearing in a set of 1960 Brooke Bond tea cards.

River pearls

Many fishes have been put to human use, beyond their values in angling and gastronomy. One such is the bleak which, perhaps surprisingly, was a core resource for the manufacture of synthetic pearls.

Right up to the 1870s, substantial commercial fisheries were located on the English rivers Trent and Thames. Bleak were harvested in their millions from the abundant populations found there. These bleak were captured for their scales, which were scraped from the flanks of captured fish. The stripped bleak, devoid of their valuable scales, were subsequently discarded or else sold on as a cheap food. In eastern Europe, by-products of this process were used for fertilizer and animal feed.

Mud and slime were cleaned from the dislodged scales, after which they were soaked in water to loosen their silvery guanine crystal pigment. Dislodged pigment was allowed to settle, accumulating at the bottom of the vessels. The silvery guanine was then used for the manufacture of the valuable product of pearl essence (*essence d'orient*). Some reports suggest that between 4000 and 5000 bleak were required to produce just 100 grams (3½ ounces) of *essence d'orient*. Demand for pearls made of this *essence* was so high that there are written records of the price of a quart (2 pints or 1.13 litres) of fish scales ranging from one to five guineas. In current terms, these 1870 prices of between £1.05 and £5.25 would equate to approximately £115 to £575 per quart today.

The finished bleak pearls were formed by mixing *essence d'orient* with wax or dense fat. However, the technology moved on with innovations in France creating far more durable bleak pearls by drawing *essence* into thin glass tubes, blown into hollow glass beads of varying forms and sizes. Some other silver fish, including roach and dace, were also used for scale production, though bleak were not only by far the most abundant species but were also the source of the best-quality *essence d'orient*.

Demand for bleak pearls was high. One of their traditional uses was to adorn the characteristic costumes of East London's cockney 'pearly kings and queens'. There were many more uses besides, including for example in earrings

and necklaces. At the height of their fashionable appeal, ornaments manufac-
tured with bleak pearls bore the name 'patent pearl'.

While the practice of manufacturing bleak pearls is occasionally reported
as still continuing in continental Europe, there is no clear supporting evidence
that this is so.

Bleak pearl manufacture was in time largely superseded by 'Roman
pearls'. These made use of nacre derived from the swim bladders of *Atherina*
(types of sand-smelt) harvested from the Mediterranean. Although British estuaries host one species of *Atherina*, the sand-smelt (*Atherina presbyter*) mentioned later in this book, no equivalent 'Roman pearl' industry was established
in Britain. Today, foreign production, particularly using plastics, dominates the
artificial pearl market.

Bleak as food

The flesh of the bleak is said to be tasty, Izaak Walton writing in his 1653 book
The Compleat Angler that,

> Bleaks be most excellent meat.

Eric Marshall-Hardy also commended the bleak, albeit with faint praise, in his
1943 *Coarse Fish*, though writing that,

> One can do worse than eat a dish of Bleak cooked after the manner of sprats, but
> they do not rank high as food.

Kenneth Mansfield was also less than fulsomely supportive of the virtues of
bleak flesh in his 1958 book *Small Fry and Bait Fish: How to Catch Them*, writing
that,

> ... though bleak were once sold from fish barrows throughout the country, they
> have passed from the modern menu. I have had sprat-sized bleak fried in oil, and
> prepared thus they make a rich and pleasant meal – but I think any other fish of
> similar size might give the same result. They were once popular soused as one
> souses or marinades a herring or mackerel.

In the days of intensive bleak fisheries to support the pearl industry, bleak from
which the scales had been stripped were said to provide cheap meat eaten by
some as well as used as livestock feed. There are though no records of the routine targeting of bleak exclusively for human food. Bleak may be taken for the
table as by-catch in some European continental freshwater fisheries, but their
use there is mainly also as animal feed.

Bleak and nature conservation

Bleak are assessed as of 'Least Concern' (LC) on the IUCN Red List (The
International Union for Conservation of Nature's Red List of Threatened

Species), reflecting a low risk of extinction since the species is abundant and widespread. Bleak also do not feature explicitly in any other British or European nature conservation legislation.

However, adverse ecological impacts have been reported in places in Mediterranean countries where, though naturally absent, bleak have formed dense populations after introduction.

A bleak tale: 'Dancing Silver'

While epic tales from feted authors about the mighty bleak have yet to be told, here is the tale 'Dancing Silver' that I published in my 2008 book *The Little Book of Little Fishes*, now sadly long out of print:

"The river was grumpy with summer heat, flows slowed to a dawdle. Bright sunlight on warm, clear water was hardly alluring to the angler, though its warmth on the skin was as pleasing as the drowsy buzz of insect life to the ears.

"I picked the spot out from some distance away. The pulse of the river skipped a quicker beat over the outcrop of limestone, brightened by crowfoot that prospered in the keener pace. A cascade fell over rock and pebble, broken water sucking in life-giving air. Then away it fell again to a cantering ripple over the deeper water beneath. In here, surely, fish life would be gathered, invigorated by oxygen and hunting amongst the profusion of insect life that adorned the mats of weed.

"I stalked, rod and body skulking from the bright horizon, finding refuge behind a sallow where the water broke from allegro to andante at the head of the run.

"Baiting quietly with impatient hands, I worked the current, enquiring of each ripple, turn and back-eddy. Again, I ran my bait through the flow, holding back to query the chaotic currents.

"Three times. Four, five times. Each time a steady, probing search.

"Sixth run down, the telltale sign! I tightened the line instinctively, to be met by a twisting, tumbling thrill on the end of the line. A flash in the summer-yellowed water signalled a fish down below the ripple, as it turned and bucked. The rod tip danced like tall verge-side grasses in a breeze, partnering the fish below in an unfettered tango.

"I let it shimmy and drive, struggling to free itself from the invisible, unknown force suddenly pulling back upon it from upstream. And slowly, steadily, it tired. And soon it was coming back to me through the ripple.

"As bright as the shards of mid-day sunlight shattered by the dimpling waters, it came to me across the surface. Skittering across the flow, periodically thrashing to hide beneath it, I was winning line.

"And soon it was to hand, a bar of silver freshly forged from the dull ore of the summer river.

"But this brilliant piece of dancing silver was no heavy metal, grown huge and strong through years feeding at sea. It was, however, no less fresh and bright, nor any less streamlined. Nor was it any less vital with living energy, nor graceful of form. But it was quite a bit smaller!

"The bleak lay on my moist hand, shimmering periodically as it sought fruitlessly to swim to freedom through yielding air. The fine hook still protruded from the corner of its sleek, upturned mouth, its fins held erect and outwards to seek purchase on the cocoon of water from which it had never before been parted. Its elegant lines, long anal fin, flaring gill covers, all attuned perfectly to quick rippled water. This was a piece of natural perfection, brilliantly crafted by billions of years of evolution to thrive in its special niche within our moving waters. And it did so here admirably, the summer river alive with hungry shards of silver that would at times boil the water's surface where a handful of loose-feed was thrown.

"This one was about two ounces but, on rare occasions, they grew to six ounces here. Many people hated them for their insatiable but, viewed dispassionately, immensely successful ways. And certainly they could be a bit of a pain at times when searching out the larger roach, dace and chub that sought streamy waters in the dour days of summer heat.

"But what, I wondered, if they went to six pound, and not six ounces? Would we then stare down our noses at such a wonder of the natural world? In truth, they would be no less wonderful whatever their size, but we would certainly notice that fact just a bit more!

"Imagine the surging run of a fish designed for speed and rippling water! Imagine the electric run and twisting power as the float was wrenched from sight in the shallow riffle! Imagine casting a dry fly for these silver powerhouses, their upturned mouths ever-eager to intercept a floating offering from the dappled surface before sending it to foam and boil as the hook drew home!

"Imagine. That's all it takes to gaze anew and in awe upon one of nature's dancing silver miracles with fresh, appreciative eyes."

Though abundant and widespread, bleak (Fig. 4.4) are none the less important elements of riverine and some still-water ecosystems with many properties to commend them to nature-lovers, anglers and anyone concerned with the vitality of the natural world.

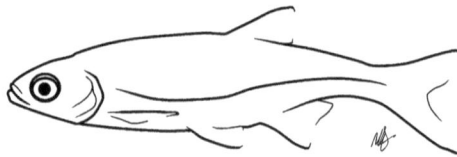

Fig. 4.4. Outline of a bleak. (Image © Mark Everard.)

Stone Loach: Stones Unturned   5

Given my childhood association with and fondness for stone loach (Fig. 5.1) – about the only species accessible to me in the small muddy streams in walking distance from where I then lived – I am amazed that more people have not indulged in the wonders of recreational stone loach angling. Seriously: they are huge fun, eager biters and offer much-neglected and free sport! While they do not so preoccupy me now, I still look on this small, shy fish, often resting in cover unseen beneath our feet, with a considerable degree of affection.

Natural History of the Stone Loach

What is a stone loach?

Stone loach, known to science by the Latin name *Barbatula barbatula* (Linnaeus, 1758), are one of the 716 members of the brook loach family Nemacheilidae found primarily in fresh waters but some in brackish waters.

Stone loach are small fish with a maximum record length of 21 centimetres (about 8¼ inches), though most loach encountered will be considerably smaller. They have elongated, almost eel-like bodies that lack scales and that are generally dull yellow-brown with irregular blotches though highly variable with habitat. Stone loach have a small, inferior (underslung) mouth surrounded by three pairs of long barbels used to sense tactile and chemical signals in the water and sediment. The head is smooth, lacking the pair of

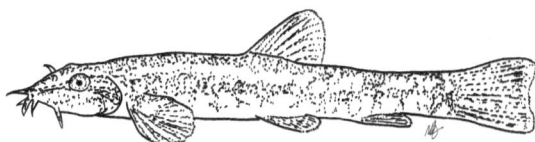

Fig. 5.1. The stone loach. (Image © Mark Everard.)

© Mark Everard 2025. *Small Fry: Britain's Tiniest Freshwater Fishes* (M. Everard)
DOI: 10.1079/9781836991700.0005

backward-pointing spines beneath the eyes characterizing the spined loach (*Cobitis taenia*); the barbels of the stone loach are also appreciably bigger than those of the spined loach.

The fins of the stone loach are rounded with a short base. The dorsal fin (III/6-8) is set to the rear of the back, as is the anal fin (III/5-6) beneath the body.

Stone loach distribution, habits, diet and senses

Stone loach are found principally in running fresh water including in small streams and, occasionally, the shorelines of lakes and some canals. They tend to be secretive by day, concealing themselves under stones and dead wood on sandy, muddy or stony bottoms, emerging to feed in half-light or full darkness. Adult fish are largely solitary, though favourable refuges may harbour other loaches as well as bullheads.

Stone loach are widely distributed across Eurasia, northwards from the Caucasus, Pyrenees and Alps as far as Sweden and Finland and eastwards through Asia as far as China. This fish occurs throughout the British mainland, except Scotland.

Stone loach feed primarily in the dark on a range of invertebrates, such as freshwater shrimps and various aquatic insects and insect larvae.

One of the more interesting features of the stone loach is that it can breathe air, taking in a small bubble to supplement its oxygen uptake from water via the gills. Whether this is obligate (essential) or facultative (an additional option) is uncertain.

Stone loach reproduction and development

Stone loach spawn in the spring, potentially spawning multiple times within a season in more productive environments. At this time, female stone loach shed clusters of sticky eggs in open water near gravel, submerged stones and plants, into which the eggs drift and adhere. Each spawning event may be brief, but individual females may release their eggs episodically over a number of days. Although female stone loach have sometimes been reported guarding the eggs, this species is generally thought not to exhibit brood protection and so these associations between female stone loach and their eggs are likely to be coincidental.

When the larvae emerge, they are benthic (living on the bed), preferring a sandy bottom and slow current. As they grow, their preference shifts to gravel bottoms and faster currents with adequate cover within which to hide by day. Habitat in which to secrete themselves is important for the growth and survival of this secretive fish.

The maximum reported age of a stone loach is 7 years.

Stone Loach and Angling

True-life confession: I used to be an avid stone loach angler when I was of primary-school age!

This was largely through necessity, lacking transport or anyone to take me anywhere, and also lacking money. But some of the neglected streams running through woods and farmland within a few miles of where I lived had good stocks of stone loach, though little else. And, with a tub of small red worms collected from piles of rotting leaves and a cane with light line tied to the end, a size 20 hook and a rudimentary float (often a bit of stick held to the line by bicycle valve rubber), a day's sport was not only possible but greatly to be enjoyed!

The water in these streams was almost invariably murky due to the clay and silt beds over which they ran and generously stocked with fallen sticks and bark under which loach abounded. 'Trotting' may be rather too grandiose a term for what I was doing, but allowing the current to roll the red worm along the silty bed beneath the crude float proved a fruitful way to attract the attentions of stone loach. Many is the happy day spent miles from anyone's gaze, sharing the seclusion with wildlife of all kinds. I recommend it!

Stone loach can also, of course, be 'hunted' in the same way as bullheads, with which they often share the same 'caves' under stones and dead wood in streams. Turning large stones and sunken pieces of timber in these places can often reveal loach. I used to find that putting a net immediately below of the 'cave' before lifting it could often result in the capture of a loach if it bolted off downstream, but I was dextrous enough back in the day to also catch a few with my hands albeit that they were far more agile quarry than the more easily caught bullheads! But the same cautionary note applies as for locating bullheads: be sure to replace the cave exactly as you found it, as this may be the home to which these little fishes are loyal.

Stone Loach and Society

The exacting water-quality needs of stone loach mean that they can serve as useful indicators of pollution, though their air-breathing capabilities mean that these fish can tolerate a degree of organic pollution and low oxygen levels.

Other names by which stone loach are known

The stone loach is also known across the British Isles by a variety of local names including simply 'loach' but also 'stoney', 'beardie' or 'groundling'.

Elsewhere across Europe, the common names applied to the stone loach include:

French	Loche, Schmerlé
Dutch	Bermpje
German	Bartgrundel, Flaßschmerle, Gäfe, Göse, Grundel, Gründel, Krasel, Lutte rümpfehen, Mös, Schmardel, Schmarling, Schmerl, Schmerle, Schmerlem, Schmerling, Schmirlin, Schmirlitt, Schmurgel, Sibirische Bartschmerle, Sibirische, Smerle, Steingrund, Zirle, Zirta
Danish	Almindelig smerling, Smerling
Swedish	Grönling

Again, the range of German names is impressively wide!

Stone loach and the arts

The stone loach appeared among a collage of other fishes painted by A.F. Lydon (Fig. 5.2) illustrating the Reverend W. Houghton's 1879 book *British Fresh-Water Fishes*. These images have been reproduced and used widely ever since.

Stone loach also featured in the series of tea cards produced by the Brooke Bond company in 1960 (Fig. 5.3), artist unknown, but since also widely appreciated and reused.

Fig. 5.2. Painting of a stone loach by the artist A.F. Lydon.

Fig. 5.3. Painting of a stone loach appearing in a set of 1960 Brooke Bond tea cards.

Stone loach as food

Few records are to be found about the gastronomic virtues of loaches. However, Kenneth Mansfield wrote in his 1958 book *Small Fry and Bait Fish: How to Catch Them* generically of the British loaches – though given the elusive nature of the spined loach this will have related principally to the stone loach – that,

> In the past loaches were considered good eating. Probably they are, but the difficulty of catching enough to make a reasonable meal has kept them off the diet sheets of all in this country except experimenters and eccentrics.

Stone loach and nature conservation

Stone loach are not an uncommon fish, though siltation of rivers and streams can compromise their abilities to find cavities under stones and woody matter in which to hide. They are assessed under the IUCN Red List (The International Union for Conservation of Nature's Red List of Threatened Species) as of 'Least Concern' (LC) in terms of extinction risk due to their wide and common occurrence.

Stone loach are not listed explicitly under other strands of British or European nature conservation guides and legislation.

Stone loach and other social connections

There are, as far as I can tell, no great works of art – painting, musical, theatrical or other – dedicated to the stone loach. Nor, as far as I can determine, has any nation, region or warrior caste selected the stone loach as its emblem!

Stone loach (Fig. 5.4), being naturally reclusive, are probably quite happy with this situation!

Fig. 5.4. Outline of a stone loach. (Image © Mark Everard.)

Spined Loach: The Cryptic Fish with a Spiny Surprise

Spined loach (Fig. 6.1) may just be the most overlooked and well-hidden native British freshwater fish species. They are highly localized in distribution in England, and they also lack any migratory behaviour. While this makes them vulnerable to poor river and drainage channel management, it also keeps them well hidden from potentially prying eyes.

Natural History of the Spined Loach

What is a spined loach?

The spined loach is known to science as *Cobitis taenia* Linnaeus, 1758. This fish is one of the 216 freshwater, and occasionally brackish-water, members of the spined loach family Cobitidae. Like all members of the Cobitidae, the spined loach is characterized by a spindle- or worm-like body that is elongated and strongly laterally compressed. The body is covered by minute scales, and the flanks have a light-brown coloration, often mottled, with 19 brown spots on each side. Spined loach are small fish, growing up to a maximum recorded length of 14 centimetres (about 5½ inches) and up to 30 grams (just over 1 ounce), though most individuals are considerably smaller than these records.

The head is small, also with a small, inferior mouth surrounded by at least three pairs of short barbels. Located in a skin pouch below and in front of each

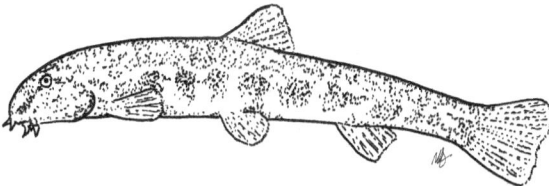

Fig. 6.1. The spined loach. (Image © Mark Everard.)

© Mark Everard 2025. *Small Fry: Britain's Tiniest Freshwater Fishes* (M. Everard)
DOI: 10.1079/9781836991700.0006

eye there is a strong, retractable double-pointed spine. The shortness of the barbels and the presence of the retractable spine in the spined loach are a key distinguishing feature from the long-whiskered, spineless stone loach.

The fins are short-based, the dorsal (III/6-8) set mid-way along the back and the anal (III/5) towards the rear of the underside. The mottling of the flanks continues into the dorsal and caudal/tail (0/15-15) fins.

One of the most unusual 'talents' of the spined loach is that these small fishes can gulp in a bubble of air to supplement oxygen absorbed by their gills. While there is no British freshwater fish that can wholly subsist by air-breathing – unlike some such as the arapaima from the Amazon, the various lungfish species from Africa and Australia, and New Zealand's Canterbury mudfish (*Neochanna burrowsius*) that has a permeable skin enabling it to absorb around 40% of its oxygen needs – this capability is a useful adaptation for the spined loach to persist when necessary in lower-oxygen conditions.

Spined loach distribution, habits, diet and senses

The spined loach occurs in freshwater rivers throughout Europe and Asia, from the Atlantic drainages from the Loire northwards to the Baltic basin (including some in lower-salinity brackish water) and eastwards to the northern Black Sea basin. This fish prefers slow-flowing rivers and backwaters including drainage ditches, favouring dense vegetation and sandy substrates. Spined loach are not highly mobile, living in dense submerged vegetation, under rocks or buried in the sand or silt throughout the day, and only becoming active to feed by night.

In Britain, spined loach are almost entirely restricted to a few eastern-flowing catchments in England: the Welland, Nene, Great Ouse, Witham and Trent. Furthermore, even within this limited range, scientific analysis reveals small genetic differences between spined loach populations found in the Welland, Nene and Great Ouse systems compared with those from the Witham and Trent. The low capacity for dispersal, weak swimming behaviour and tendency to remain inactive may contribute to this relatively localized genetic diversity.

Spined loach fare less well in open channels that lack dense habit in which to hide from competitors and potential predators, but marginal backwaters, densely vegetated channels and channel edges transitioning into connected pools and wetlands, tributary streams and ditches can all provide suitable habitat. Habitat variety is in fact vital for the survival of these fish, which depend on ditches and lowland drainage channels that are not over-managed, and rivers that are not excessively modified by urban, industrial and agricultural encroachment. Well-vegetated canals that have become abandoned, barely or no longer used for navigation, can also provide valuable habitat.

Given both the small size of this fish and its tiny mouth, most of the diet is derived by filtering sandy substrates for fine bottom-living invertebrate as well as some vegetable food particles. The chemically sensitive barbels help this fish locate suitable food items.

Spined loach reproduction and development

Spined loach spawn in spring, parent fish swimming excitedly and syn- chronizing movements until the male fish entwines itself around the female, squeezing her and triggering the release of eggs. The sticky eggs are depos- ited in unguarded clusters on submerged plants, roots or stones. The hatching larvae are tiny, hiding under vegetation and in other cover until they become free-swimming and can start exogenous feeding. They require dense vegeta- tion or other cover for their continued survival right up to adulthood. Spined loach are reported as living up to 5 years.

Another of the more fascinating features of the spined loach is that it has been reported that gynogenesis can occur. Gynogenesis is a form of partheno- genesis or, in other words, a system of asexual reproduction though one that requires the presence of sperm to trigger egg development. However, the sperm, which need not come from the same species, does not fuse with the genetic ma- terial in the nucleus of the egg. Gynogenesis is consequently often referred to as 'sperm parasitism', given the fruitless role of male gametes. The extent to which gynogenesis naturally occurs in spined loach populations is not known, though it may be an adaptation to the fragmented distribution and lack of mo- bility of these fish. A further unsolved mystery is the provenance of the sperm triggering gynogenesis in spined loach; it is known, as we will see later in this book, that sperm of a range of other fish species suffices to trigger this process of asexual reproduction in gibel (*Carassius gibelio*).

Spined Loach and Society

I feel I have to start this section with an apology to the spined loach. This fish is far more elusive and scattered in distribution compared with its cousin, the stone loach, and far less is written about it. However, there are several inter- esting aspects of the spined loach. These include its nature conservation sig- nificance, its potential as a pet (at least historically) and its diminutive size. By and large, the spined loach lives in the shadow of its cousin, with some gen- eral remarks common between the two unrelated loach species. But, as one of nature's recluses, this may not unduly perturb this shy little fish!

Other names by which spined loach are known

The spined loach is alternatively known as the 'spiney' but also, like the stone loach, as simply the 'loach' or 'groundling'.

Elsewhere across Europe, the common names applied to the spined loach include:

French	Dorngrunder, Loche de rivière
Dutch	Kleine modderkruiper
German	Dorngrundel, Sandputtler, Steinbeiß, Steinbeisser/Steinbeißer, Steinpeitzge, Steinschmerl
Danish	Almindelig pigsmerling, Pigsmerling
Swedish	Nissöga

It is quite clear that the British are less adventurous with bespoke names for the spined loach, most of our common names being shared with the stone loach.

Spined loach and the arts

The spined loach was another species depicted in a collage of other fishes painted by A.F. Lydon (Fig. 6.2), illustrating the Reverend W. Houghton's 1879 book *British Fresh-Water Fishes* and reproduced and used widely ever since.

Spined loach were also featured in the series of tea cards by the Brooke Bond company in 1960 (Fig. 6.3), images that have since been widely appreciated and reused.

Fig. 6.2. Painting of a spined loach by the artist A.F. Lydon.

Fig. 6.3. Painting of a spined loach appearing in a set of 1960 Brooke Bond tea cards.

Spined loach as pets

While many of Britain's freshwater fish species grow too large for the everyday home aquarium, this is not so for the spined loach. However, as we will see, there are legal restrictions on the 'exploitation' of this species which make its retention in captivity unacceptable. Also, given the exacting requirements of this small fish for filter feeding fine edible particles from sandy and silty surfaces, keeping them adequately nourished may present a challenge.

Another member of the spined loach family Cobitidiae, however, was once kept in indoor aquaria, sharing the habit of the spined loach of periodically rising to the surface to ingest a bubble of air. This fish was a near relative, the European weatherfish (*Misgurnus fossilis*), though this species does not occur naturally in Britain. Like the spined loach, the European weatherfish can utilize oxygen from the air if dissolved concentrations in the water become depleted. It is this habit of coming to the surface to gulp air, occurring more frequently under different climatic conditions, that not only gave the weatherfish its common name but was also the principal attraction of keeping the fish in the home as a kind of living barometer.

I have visited many countries and, like here in Britian, I find it perennially baffling that the attractive fishes found in those places are not valued among fishkeeping hobbyists, who prefer instead alternative gaudy residents of foreign lands.

Spined loach and nature conservation

Spined loach are locally distributed, with threats to their naturally fragmented populations posed by over-management of waterways removing the dense vegetation that they require for cover. They are listed under various strands of nature conservation guides and legislation including:

- The spined loach is assessed under the IUCN Red List (The International Union for Conservation of Nature's Red List of Threatened Species) as of 'Least Concern' (LC) in terms of extinction risk across its broad Eurasian range, despite its restricted distribution in English waters.
- The Bern Convention on the Conservation of European Wildlife and Natural Habitats 1979 (generally referred to as the Bern Convention or Berne Convention) lists the spined loach in Appendix III, which is concerned with species for which exploitation is controlled.
- The European Union (EU) Habitats Directive (Council Directive 92/43/EEC on the Conservation of Natural Habitats and of Wild Fauna and Flora) lists spined loach in Annex II, comprising animal and plant species of community interest whose conservation requires the designation of Special Areas of Conservation (SACs).
- The UK Biodiversity Action Plan (UK BAP), the UK Government's response to the Convention on Biological Diversity (CBD), lists the spined loach in England only (not in Scotland, Wales or Northern Ireland).

Not the smallest but perhaps the rarest native British freshwater fish

Although the shy and elusive spined loach is sometimes claimed to be the smallest native British freshwater fish, with a maximum body length of up to 14 centimetres (about 5½ inches), it is substantially bigger than the three-spined stickleback's 11 centimetres (almost 4½ inches) from the front of the head to the tail. However, it is the ten-spined stickleback that claims the 'smallest native British freshwater fish' record at just 9 centimetres (3½ inches) from pointy snout to tail.

But if not the smallest, the spined loach is most likely the scarcest small fish in British fresh waters. Based on its highly localized distribution, lack of mobility, small size, tendency to remain hidden in dense cover and its vulnerability to over-management of vegetated drains and backwaters, the spined loach almost certainly qualifies as Britain's rarest fish species (Fig. 6.4).

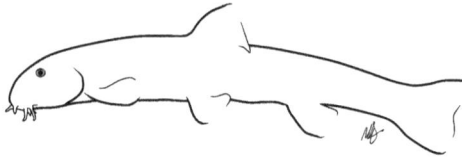

Fig. 6.4. Outline of a spined loach. (Image © Mark Everard.)

Bullhead: The Freshwater Troglodyte

Beneath stones and logs in your local stream may lurk a large-mouthed, frog-eyed predator, ready to pounce at any moment on unwary prey. All large head, spiky behind with a tapering body, pressed hard to the stream bed, they wait. Dare you venture unguarded into such precarious waters? Only with extreme caution, if you were the size of a fish fry or a large invertebrate!

Bullheads (Fig. 7.1) are quite unlike other fishes found in the fresh waters of Britain. With small eyes situated high on a disproportionately large head, gaping mouth and smooth, tapering body, I am minded of a diminutive angler fish. This is perhaps not altogether surprising as they are freshwater members of the sculpin family, many or most of which inhabit marine environments, lending this charismatic if diminutive fish its unique features. Most *Cottus* species live in fresh waters, but a few inhabit coastal marine environments. Given their relative immobility, *Cottus* species tend to live cryptically with a high degree of loyalty to suitable habitat. Consequently, they are prone to a significant degree of speciation.

Natural History of the Bullhead

What is a bullhead?

The bullhead – Latin name *Cottus perifretum* Freyhof, Kottelat and Nolte, 2005 – is the only British species of the sculpin family (Cottidae), the 289 species

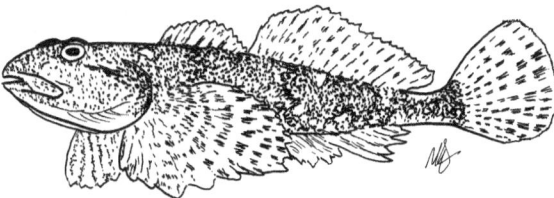

Fig. 7.1. The bullhead. (Image © Mark Everard.)

© Mark Everard 2025. *Small Fry: Britain's Tiniest Freshwater Fishes* (M. Everard)
DOI: 10.1079/9781836991700.0007

of which occur in marine, brackish and fresh waters. These fishes have a scaleless body armed with prickles, and they all lack a swim bladder (an air-containing organ found in many other fish species enabling them to adjust their buoyancy).

Some readers may question use of the Latin name *C. perifretum*, having known the species by the name *Cottus gobio* assigned in 1758 by Carl von Linné. The name *C. gobio* was attached to the fish for 247 years until it was reclassified in 2005. The story of the reclassification, for those interested to know more, is outlined in Box 7.1.

Bullheads in Britain, found only in fresh and mildly brackish waters, are small fish with a maximum recorded length of 18 centimetres (7 inches) and a weight of up 20 grams (less than 1 ounce), though they are commonly much smaller. The maximum reported age for bullheads is 10 years.

The head of the bullhead is large and broad, the eyes small and located on the top of the head, with a big, terminal mouth that lacks teeth, these features contributing to the common name 'bullhead'. The gill cover of the bullhead is armed with prickles. Behind this, the smooth and scaleless body tapers away, mottled brown on the back and flattened beneath adapted to a life on the bed. The lack of a swim bladder is a further adaptation to the exclusively bottom-dwelling habit of these small fishes.

Box 7.1. Reclassification of the bullhead found in Britain as *Cottus perifretum*

All freshwater bullheads found across Europe were formerly thought to have been the single species *Cottus gobio*, described by Linnaeus in 1758. However, given the relative immobility of these small fishes and their fragmented distribution even within single drainage basins, separation into distinct genetic lineages was likely.

A taxonomic revision in 2005 by Freyhof *et al.* (the full reference is given in the Bibliography at the end of this book), based on molecular studies integrated with morphological characters, recognized 15 distinct *Cottus* species. All are assumed to have diversified from *C. gobio* into distinct morphs and species since the Pliocene epoch (5.3–2.6 million years ago).

Cottus perifretum is one of the newly recognized *Cottus* species, noted as being distributed from the Rhine river drainage in Germany, the Netherlands and France, in the rivers of Great Britain, and in drainages from the Garonne to the Scheldt and Meuse in France and Belgium, and the Mosel and Sieg rivers in Germany.

Although Freyhof *et al.* (2005) record that *C. perifretum* is the sole species thus far determined in British rivers, they also note differences in the morphology and genetics of bullheads from the River Wharfe northwards and in Wales compared with those from southern Britain and the Scheldt in continental Europe, and that "It cannot be excluded, that more than one species occurs in Great Britain".

A simplified telling of this reclassification can be found in an article in the magazine *British Wildlife* by Everard and Pickett (2025), also listed in full in the Bibliography.

Bullheads have broad, mottled pectoral fins matching the drab coloration of the body. There are two dorsal fins, the front dorsal fin held aloft by short spines (VI-VIII) and the rear dorsal is supported by 15–18 soft rays. There is a long anal fin lacking spines (0/10-13), and the caudal (tail) fin is supported by 13 or 14 soft rays (0/13-14).

Bullhead distribution, habits, diet and senses

Bullheads are fishes of well-aerated water containing rocks or woody matter, including logs and tree roots, beneath which they can live out their whole adult lives largely hidden from view as troglodytes – cave dwellers – remaining loyal to the same territorial 'cave'. Unlike virtually all other British freshwater fishes, except spined loach, bullhead seem to exhibit no migratory behaviour.

Bullheads are common in the gravel or rocky bed of rivers, wave-lapped edges of larger still waters and some mildly brackish waters, favouring well-oxygenated waters throughout mainland Britain. However, they cannot move between isolated river catchments without assistance as this fish cannot tolerate fully saline water.

A range of bullhead species occurs across Europe from the streams and shores of the Baltic Sea, eastwards and southwards across the major drainage basins of northern Europe. However, *C. perifretum* is restricted just to British fresh waters and river drainage basins predominantly in Germany, the Netherlands and France. These continental basins include the Rhine (Germany), Meuse (a major river of France and Belgium) and the Scheldt (spanning northern France, western Belgium and the south-western part of the Netherlands) that share a common delta in the Netherlands, as well as the Mosel and Sieg rivers (Germany) and the Garonne (south-west France with some headwaters in northern Spain).

The bullhead's natural distribution across Britain was restricted to historic connections of British rivers to continental catchments through the 'Doggerland Bridge' before its collapse at the end of the last ice age (between 6500 and 6200 BC). However, like many freshwater fish species, bullheads have since been more widely distributed and naturalized across the British Isles, largely as an unintended stowaway when other fish or water plants have been translocated. Bullheads through are still absent from Ireland, which lacks a land or river bridge with mainland Britain.

The bullhead, along with the brown trout (*Salmo trutta*), is also a fish that can adapt to living in complete darkness in streams running through some cave systems, particularly in localities in Wales where bullheads occur in underground streams all year round. Bullheads living in caves seem completely unaffected by their dark surroundings, their body colour not seeming to change.

Bullheads are exclusively carnivorous, feeding on invertebrates and fish fry.

Bullhead reproduction and development

The male bullhead takes charge of parental duties. Bullheads spawn in Britain typically between March and May. At this time, male fish display to generally smaller females living in adjacent territories. If a female is receptive, she may venture into the lair of the courting male, depositing a clump of around 100 sticky eggs on the cave roof. Bullhead eggs are typically 2 to 2.5 millimetres (around 0.08 to 0.1 inches) in diameter and amber in colour. Once the eggs are laid and fertilized, the male bullhead drives the female away. He then guards the eggs for between three and four weeks until they hatch, continuing to protect the hatchlings for ten or 12 days while they consume their yolk sacs. After this time, juvenile bullheads become free-living, dispersing to seek the own territorial shelters and to evade the potential predatory attentions of their fathers. As bullheads grow, they feed on sequentially larger invertebrates and fish fry.

There may be multiple spawnings during a fair late spring and early summer season.

Bullheads as prey and predator

Bullheads can serve as 'bite-sized snacks' for a wide variety of piscivorous (fish-eating) birds, fish and mammals. Kingfishers, egrets and herons will gladly opportunistically accept a bullhead if one emerges from its cave.

Watching the feeding behaviour of the Eurasian otter (*Lutra lutra*), a 'stone turner', gives insight of why surveys of otter spraints (droppings left as territorial markers) reveal that bullheads can comprise as much as 90% of the diet of this mammal. Otters do not tend to waste energy chasing down active fishes but, rather, using their stout sensitive whiskers, they instead probe around and beneath submerged rocks, debris and in silt for more sessile, hidden prey – significantly including European eels and bullheads as well as loach. The introduction of invasive American signal crayfish (*Pacifastacus leniusculus*) into Britain in the 1960s and their subsequent establishment and spread now mean that these large and destructive invertebrates can constitute a significant element of the diet of the otter where present, evidenced by large amounts of crayfish remains in their spraints. While the presence of signal crayfish is not welcome, this may give the bullhead some relief as otter populations recover from their near extinction in England by the 1980s due to bioaccumulation of persistent synthetic pesticides.

Bullheads are exclusively predatory, feeding on invertebrates and fish fry. It is this habit that has made the species unpopular with some fishery managers, this small fish allegedly preying on Atlantic salmon (*Salmo salar*) and brown trout eggs and fry. It is undoubtedly true that, as opportunistic and voracious predators, they may take some where the species coexist, though the extent to which this impacts salmon populations is far from certain, with bullheads thriving on a largely invertebrate diet – freshwater shrimps and other crustaceans, insect larvae and other invertebrates – in waters in which salmonid fishes are not present. Though once cast as a villain, many river managers today have a far more sympathetic perspective about the presence of bullheads as indicators of a healthy river system, better informed about the relative contributions of predation versus the many more insidious human-induced pressures on our river systems.

Bullheads and Angling

The bullhead is one of many small fry species that I grew up with, visiting local streams and turning over stones to find and trap these charismatic little fishes, all frog-eyed and mouthed, in my bare hands. Decades later, I did so with my daughter too, including for the camera when presenting on the BBC TV *Springwatch* programme. Huge fun is to be had at the mere inconvenience of chilly hands and, often, overtopped wellington boots as the thrill of the chase overrides caution!

It is only fair at this point to remind people to respect the fact that bullheads are loyal to stones or other caves, potentially living out their whole lives under the same rock, wooden debris or other niche. This 'cave' offers them shelter from would-be predators and strong currents, food in the shape of invertebrates colonizing the rock or settling in the slack water around it, and of course a place to nest and nurture young. Consequently, if you turn over stones or woody debris to find bullheads, remember always to turn them back again exactly as you found them. Also, return any fish you catch to exactly the same spot. Without so doing, the bullhead will be disoriented, and potentially even separated from its brood of eggs.

From time to time as a kid, I also presented scraps of worm on tiny hooks around the edges of likely-looking stones or woody matter. This is specialist angling for sure but was and remains great fun! Perhaps I should do it more often.

More than once over recent years, I have caught bullheads while fishing for roach with a bread-filled swimfeeder. On these occasions, the quivertip on my leger rod has jagged minutely, and I have hit the bite immediately knowing what delicate feeders winter roach can be. Feeling no resistance, I have reeled in to change the bait and recast but been amazed to find that a bullhead has firmly wedged itself in the swimfeeder, bracing itself against the plastic mesh by flaring its broad pectoral fins. I guess that, down in the depths, the bullheads have seen the cage feeder as a suitable cave to explore, jutting out their broad fins on finding themselves elevated from the riverbed and into fresh air. On these occasions, I am not sure which of us – me or the bullhead – has been more surprised!

Bullheads and Society

Other names by which bullheads are known

Bullheads go by a wide variety of local names. One of the most common of
these is the 'miller's thumb', so named as bullheads were common in the races
of mills. The broad head and tapering body of this little fish was thought to be
similar to the broadened thumb of millers, who spent years testing the con-
sistency of their milled flour by pressing it with their thumbs. The Reverend
W. Houghton comments in his 1879 book *British Fresh-Water Fishes*,

> The Miller's Thumb is supposed to resemble that organ in the miller, which is said
> to assume a flattened form from frequently testing the flour.

A small subset of alternative local names across Britain includes 'bullyhead',
'mullyhead' and 'wayne'.

In France, *C. perifretum* is known as the 'chabot fluviatile', which trans-
lates as the 'fluvial sculpin'.

Elsewhere across Europe, the common names applied to bullhead species
(which, let's recall, are probably mainly not *C. perifretum* in the light of the
2005 reclassification) include:

French	Botte, Cabot, Chabot, Têtard (among others)
Dutch	Rivier-donderpad
German	Dickkopf, Dickkopp, Dolbn, Greppe, Groppe, Kappen, Kaulkopf, Kaulquappe, Kautzenkopf (among others)
Danish	Almindelig ferskvandsulk, Ferskvandsulk, Flodulk, Hvidfinnet ferskvandsulk
Swedish	Stensimpa

The alternative French name têtard actually means 'tadpole', reflective of the
shape of this fish. Likewise, the alternative German name Kaulquappe also
means 'tadpole'.

The Latin name of the bullhead is also of interest. The genus name *Cottus*
is said (by Scharpf, 2024: see the Bibliography for the full reference) to be a
Latinization of the Greek *kottos* (the original form *koviós* or *kóthos*), roughly
translating to 'head' and applied to small fish with a large head, and so used
for the sculpins.

The specific name *perifretum* combines the prefix 'peri-' (around) and
the Latin 'fretum' (meaning 'strait', *Fretum Gallicum* referring to the English
Channel) as the species inhabits streams on both sides of the English Channel.

Bullheads and the arts

The bullhead (albeit with a slightly annoyed look on its face!) featured in
a collage of other fishes painted by A.F. Lydon (Fig. 7.2), illustrating the

Fig. 7.2. Painting of a bullhead by the artist A.F. Lydon.

Fig. 7.3. Painting of a bullhead appearing in a set of 1960 Brooke Bond tea cards.

Reverend W. Houghton's 1879 book *British Fresh-Water Fishes* and reproduced and used widely ever since.

The bullhead also featured in the series of tea cards by the Brooke Bond company distributed in 1960 (Fig. 7.3), these images since widely appreciated and reused.

Bullheads as food

I am happy to report that I can find no evidence of mass harvesting of bullheads for human consumption! Given their scattered distribution, any such harvesting would not only be onerous but might seriously undermine the integrity of populations that might have a unique genetic heritage.

However, John Harrington Keene wrote in his 1881 book *The Practical Fisherman* that he had tried bullhead more as a matter of experimentation, describing the white cooked flesh as a gastronomic delicacy.

The Reverend W. Houghton writes in his 1879 book *British Fresh-Water Fishes* that,

> I am told by one or two persons who have eaten this fish that it is very good indeed; when the head is excluded, however, there is but little left to eat.

These are small fish, and a lot of effort would be required to collect sufficient for a mouthful, let alone how to deal with their spines. In my view, they are best left alone to live out their elusive troglodyte existence in peace!

Bullheads and nature conservation

Assessing its global risk of extinction, the IUCN Red List (The International Union for Conservation of Nature's Red List of Threatened Species) classifies the bullhead *C. perifretum* as of 'Least Concern' (LC), reflecting its broad distribution, also noting that population trends are not known.

At European scale, the bullhead is scheduled under the European Union (EU) Habitats Directive Annex II, 'Animal and plant species of community interest whose conservation requires the designation of Special Areas of Conservation (SAC)', reflecting its patchy distribution. However, this designation is based on *C. gobio*, prior to its reclassification into several distinct species. There may be a need to revise the Annex in the light of the reclassification, reflecting the vulnerabilities of *C. perifretum* and other newly distinguished species.

Bullhead tales

In my 2008 book, *The Little Book of Little Fishes*, I mused about the bullhead and the alternative realities that would come to play if this fish was much bigger:

> "If they reached five pounds, we'd all think twice about wading. If they weighed in at ten pounds, we'd keep our kids back from the water's edge. They lurk, small eyes high on their bulbous heads, a spine arming their gill covers, and their body dark, scale-less and disproportionately small. They lurk, ever poised to pounce from their underwater dens to engulf in cavernous mouths almost anything of a swallowable size that moves close enough.

> "But they are not ten pounds in weight. Not five either. They don't get to a pound, or even a half-pound. Not even a quarter. One ounce is a good one, two ounces tops. And, far from beast of the deep, they are charming little characters!

> "Just imagine what sport they would offer if they weighted a lot more! What if there really were double-figure specimens? Imagine the skill in pitting one's wit against a bulky predator that lurked, lunging with sudden and unexpected ferocity from under stones, submerged wood and dense vegetation! Visualise the thrill and energy of the 'hook and hold' fight as they gave their all to regain their refuge! Imagine how tasty might be the monkfish-like body when the fish was

banked! But let's hold on just one moment before the adrenaline-rush of all this speculation pushes us from fantasy to insanity!

"Imagine also the foot-biting, hand-gulping, child-scaring reality of large specimens of such a formidable sporting fish, lurking ready to lunge at and engulf anything that moves in front of their riverside caves! On second thoughts, let's just be happy that nature has made them as intended. Big-headed, beady-eyed, gape-mouthed and charming. ... but above all little!"

A British freshwater fish unlike others

The bullhead is a curio. It is a freshwater member of a predominantly marine lineage with some recognized nature conservation importance, and it is also a diligent father (Fig. 7.4). This is certainly a little fish with a big personality.

Fig. 7.4. Outline of a bullhead. (Image © Mark Everard.)

Three-Spined Sticklebacks: Knights in Armour

8

I grew up with three-spined sticklebacks (Fig. 8.1). Not literally, obviously, as I can't breathe underwater. But one of the formerly abundant – now sadly massively depleted – small farm ponds scattered across the landscape around where I lived as an infant was a haven for three-spined sticklebacks. Many is the time I would find myself there, smeared in cloying yellow Wealden clay from the pond's edge in my enthusiasm, pursuing these odd little fishes during what seemed like endless spring and summer days.

I loved, and love still, their stiff-bodied antics, their chain-mail armouring constraining any more lithe motion. They dart and pause, hovering with quivering fins, then dart again. These little fishes were instantly recognizable then, as they are now, when spotted from a distance. Their long faces are like the horses ridden by knights in a chess set, or perhaps little water-dragons, terminating in a small mouth, reminding me also of the equally equine features of a sea horse.

The biggest prize was the capture of a male in full breeding regalia. In breeding season, male three-spined sticklebacks morph from an otherwise unremarkable base colour of greenish-brown fading to silvery, making a dramatic transformation into veritable popinjays of the fishy world. In their finery, they are among the most exquisite of fish that swim in our waters. The flamboyant coloration of the breeding male, metallic emerald and blue from the head and eyes and across the back with a fiery red on the underside, is proudly displayed to woo would-be female partners to the nest he has constructed as their bridal

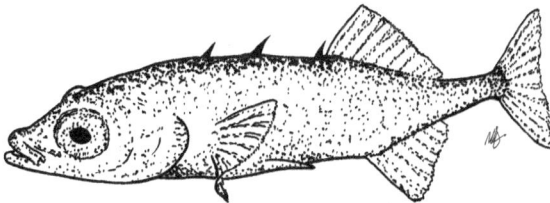

Fig. 8.1. The three-spined stickleback. (Image © Mark Everard.)

© Mark Everard 2025. *Small Fry: Britain's Tiniest Freshwater Fishes* (M. Everard)
DOI: 10.1079/9781836991700.0008

bed. Not only that, but three-spined sticklebacks prove to be 'new men', male fish defending and nurturing the eggs and the hatchlings until the fry are able to venture out on their own. What's not to like?

Add to this their hardiness and fearlessness, and their general tough demeanour. They are also perfect temperate-aquarium occupants, hardy enough to withstand poorer water quality more than other British freshwater species. In fact, they often abound in polluted urban streams, rivulets, canals and pools. For that reason, they live perhaps in closer proximity to urban children than other fish species and may have featured in formative experiences for many of us. They piqued our curiosity and admiration then and, for many of us, still do so as we fail to grow up entirely!

Natural History of the Three-Spined Stickleback

What is a three-spined stickleback?

Three-spined sticklebacks are adaptable and hardly little fellows, going by the Latin name of *Gasterosteus aculeatus* Linnaeus, 1758. They are members of the stickleback and tubesnout family (Gasterosteidae), along with two other British species, one of which is the freshwater ten-spined stickleback (*Pungitius pungitius*) and the other the marine fifteen-spined stickleback (*Spinachia spinachia*).

Three-spined sticklebacks can reach a maximum length of 11 centimetres (about 4¼ inches), though many are far smaller than this, with a maximum reported age of 8 years. The head is elongated, tapering to a thin snout and a small mouth that lacks barbels. The body is elongated, laterally compressed and held stiffly. Three-spined sticklebacks lack a covering of small scales across the body, but a few large and prominent bony plates known as scutes, or lateral plates, arm the body contributing to the stiffness of the fish. The number of scutes is variable, typically between one and over 20, lining the flanks from just behind the head extending back to the beginning of the caudal peduncle. A further seven or eight small plates, or platelets, may also be found along the peduncle rearwards to the beginning of the caudal (tail) fin. In marine populations – this is a species able to live in and move between fresh and salty water – these fish tend to be fully plated with up to 37 scutes on each flank. In fact, the Latin name of the genus *Gasterosteus* means 'bony stomach', largely referring to the line of bony scutes that generally extends along the flanks. Historically, different scute patterns were used to distinguish different 'species', now all combined into the single species *G. aculeatus* (see Box 8.1).

There are two dorsal fins, the front one comprising three but sometimes four stout spines with a small membrane on the hind edge of each. These fearsome, erectile spines deter predators, also giving the species its common name. The second dorsal fin behind the front fin is supported by between ten and 14 soft branched rays (0/10-14). On the belly, the pelvic fins are reduced to a pair

---

Box 8.1. Formerly recognized three-spined stickleback 'species' distinguished by dorsal spine, scute and plate patterns

The Reverend W. Houghton's 1879 book *British Fresh-Water Fishes* lists an additional two species of stickleback: the four-spined stickleback (assigned the name *Gasterosteus spinulosus*) and the short-spined stickleback (described as *Gasterosteus brachycentrus*), distinguished largely by the spines in the dorsal fin. Both of these reported 'species' are now regarded as variants of *Gasterosteus aculeatus*, which is known to vary in form depending on environmental factors.

In general, three-spined sticklebacks inhabiting marine environments tend to be extensively armoured with scutes and plates on the caudal peduncle, and this is known as the *trachurus* form (previously identified as a separate species *Gasterosteus trachurus* Cuvier, 1829: the rough-tailed stickleback). By contrast, three-spined sticklebacks inhabiting freshwater environments generally lack plates on the caudal peduncle and are known as the *gymnurus* or *leiurus* forms (formerly classified as *Gasterosteus gymnurus* Cuvier, 1829 or *Gasterosteus leiurus* Cuvier, 1829: the smooth-tailed stickleback). Houghton also lists two intermediate forms – *Gasterosteus semiarmatus* and *Gasterosteus semiloricatus* – though acknowledges that all four of these formerly classified 'species' are variants of *G. aculeatus*. It is not known whether these forms have evolved as an adaptation to their environment, or if they develop from common genetic stock in response to exposure to different salinities.

Comparison of European sticklebacks from high northern latitudes, where freshwater populations maintain connectivity to marine populations, with those lacking connection to marine populations in southern Europe reveals that southern populations exhibit lower genetic diversity (see Coll-Costa *et al.*, 2024, listed in full in the Bibliography at the end of this book).

---

of single sharp 'thorns' and a small soft ray. The anal fin has a single leading spine followed by between seven and 11 soft rays (I/7-11).

For most of the year, the basic body colour is mottled brown or greenish, fading to silvery or pale yellow-white on the belly. However, as we have described, the male takes on a vivid hue in the breeding season.

Three-spined stickleback distribution, habits and diet

Three-spined sticklebacks are hardy. They are commonly encountered in still waters, canals and ditches, as well as the slower-moving margins of rivers, and also occur in estuaries and even coastal waters across a wide circumarctic and temperate distribution in the northern hemisphere. They occur as far south as the Black Sea, southern Italy, the Iberian Peninsula, North Africa, in Eastern Asia to the north of Japan as well as in North America and Greenland. The hardiness and adaptability of the three-spined stickleback mean that, second only to the Arctic charr (*Salvelinus alpinus*), it has the most northerly global distribution of any freshwater fish species.

In some spate rivers, such as those found in Scotland, three-spined stickle-backs can be flushed out of the river systems to overwinter in estuaries or coastal waters, reinvading the rivers as flows abate in the spring and summer. In the south of England, these fish generally forego this marine phase, though they tolerate and can prosper in brackish waters. This tolerance of fully marine water means that they are naturally present in the island of Ireland, listed as such by the medieval clergyman and chronicler Giraldus Cambrensis in his *The History and Topography of Ireland*. These fish also have a pronounced tolerance of significant pollution, and also thrive in overgrown or shallow waters, allowing them to form dense shoals where their predators do not prosper.

The three-spined stickleback is so adaptable that it is the most widespread of freshwater fish species in the British Isles, extending from the tip of Scotland and its outer islands in the north then southwards throughout England, Wales and all of Ireland except the extreme north-west. Three-spined sticklebacks have even been found inhabiting underground streams in caves, though these sightings are generally considered to be occasional visitors rather than fully resident populations.

Three-spined sticklebacks have an entirely predatory diet, feeding on small invertebrates, fish fry and the smaller larvae of amphibians. The presence of three-spined stickleback in ponds that are important for newt populations is generally thought to be adverse for amphibian survival.

Three-spined stickleback reproduction and development

In the extended breeding season of the three-spined stickleback during the spring and early summer, the male fish develops a truly stunning coloration. Males at this time transform from an otherwise predominantly drab and silvery body colour. The upper part of the head and back and the flanks turn a brilliant metallic emerald colour, the eye becomes electric blue, and the breast and underside of the head and flanks take on a vivid red hue. By contrast, female three-spined sticklebacks retain their browny-silver colour.

It is the male stickleback that takes on parental duties. Their parental care is a wonderful thing to watch, and these fish can and do breed readily in captivity. The male fish initiates the spawning process, selecting a territory in the generally shallower margins of the streams, ponds, canals, ditches or estuaries that they inhabit, and guarding it against rival males as well as driving away substantially larger fish species. Within this territory, the male stickleback then proceeds to build a nest of vegetation, comprising pieces of water plants, fallen twigs, algae and other organic matter, glued together using secretions from their kidneys rich in glue protein known as spigin. These nests are rounded, with a narrow passage through their centre.

Once the nest is constructed, the male fish then seeks to court female sticklebacks while warding off other fish including species that may be substantially bigger. The courting dance of the male stickleback is a jagged zigzag dance, their already stiff bodies held rigid. If the female is sufficiently impressed, she will enter the tunnel in the nest and deposit anything from 100 to 400 eggs, each 1.5 to 1.9 millimetres (around 0.06 to just over 0.07 inches) in diameter. These eggs are then fertilized by the male, which then drives off the female. The courtship ritual begins again with the male attracting typically three or four female sticklebacks to add their eggs to his growing clutch, potentially leading to up to 1000 eggs in total laid within the nest.

Male sticklebacks then become devoted fathers, tending the eggs by fanning oxygenated water over them and picking out any that turn white with fungus. Fry are cared for after the eggs hatch. They are also herded and guarded for a short while after they emerge as free-swimming juveniles. After these early days, the juveniles disperse to live independent lives.

Three-spined stickleback as prey and predator

Three-spined sticklebacks are small fish, and so consequently can fall prey to a wide variety of fish, bird, mammal and other predators. The three thick 'thorns' on the front dorsal fin and the two strong spines of the pelvic fins may deter some predators or cause them to spit the stickleback out as they are hard to swallow. However, many a stickleback is taken whole as the spines are compressed to the body when the fish is swallowed head first.

Stickleback fry are prey for predatory insects such as dragonfly larvae, water beetles and their larvae, water boatmen and water scorpions, which may also take smaller adult sticklebacks.

Three-spined sticklebacks are though also exclusively predatory, feeding on living prey fine enough to ingest with their small terminal mouths. Suitable prey items include any of a range of invertebrates, fish fry and amphibian larvae.

Three-Spined Sticklebacks and Angling

Beginnings

I have surprisingly vivid memories from when I was only 2 years old and onwards into my infant years of visits to a local dewpond that was close to where I then lived in Kent. The pond was at that time on the edge of woodland bordering a field, from which cattle came to drink. This complex of tree cover and open bank, encroaching sallows and poached clay, was home to many three-spined sticklebacks. Also, a sole tench that ghosted between cover from time to time as a mysterious shadow. The sticklebacks were always eager to snatch a bait.

This pond was, in reality, tiny. It certainly appeared so when the woodland was razed to make way for a bypass, exposing the pool above the road embankment where it progressively filled with silt, neglected and unloved. But, for me in those distant, golden days, it was a vast ocean of promise and mystery. For all the pond's diminutive size, it teemed with three-spined sticklebacks. This was where the fishing bug really bit hard and has never let up its mole-like grip!

Oftentimes, with a yard of black cotton (one Christmas, I asked for a reel of black cotton of my very own and was overjoyed when I unwrapped one!) and a hazel or willow switch snapped off from a hedgerow or copse en route, I would literally tie a worm amidships on to the end of the line – no hooks required! – and dap this into the murky depths. In no time, a stickleback would gorge itself on one of the ends of the worm, not letting go as it was swung to an eager young hand to be admired, perhaps retained a while in a clay depression left as the cows came to drink, before being released back into the wild. The biggest prize was a male in full spawning livery, brilliant red on the chest with vivid blues on the head and back. (In retrospect, perhaps retaining gaudy males for any length of time was not a great idea if this distracted them from parental duties!)

Sticklebacks have retained a place in my heart ever since. Watching their staccato movements in the water's edge is a reminder of those magical times and the lifelong connections they have forged for me with the wonders of the aquatic realm.

Tactics for three-spined stickleback fishing

There are no specialist rigs or baits dedicated to this charismatic fish. In fact, little has changed in approach since my childhood pursuits other than using a more 'grown-up' rod and reel, light float or free line, often fishing by sight rather than waiting for another form of indication. Tiny hooks are a must, given the small aperture of the mouths of these fish, though oftentimes they are superfluous as sticklebacks are so reluctant to let go of a meal. Suitable baits include a small worm or fragment of a larger one, or a small maggot, as these are predatory fish disinterested in bread or other plant-based baits.

The only formula is to enjoy it, ignore people who look at you strangely for this apparently pointless pursuit, and reconnect a little with that childish glee that got us into angling and fuelled our fascination with the underwater world in the first place!

Three-Spined Sticklebacks and Society

Other names by which three-spined sticklebacks are known

Three-spined sticklebacks are known by a wide range of local common names across the English-speaking nations within their broad northern hemisphere range of this species. These common names include 'prickleback', 'sticky-bag', 'tittle-bats', 'tiddlebat', 'banstickle', 'barstickle', 'barnstickle', 'barny-stickle', 'barnytickle', 'branchy', 'branstickle', 'burnstickle', 'baggie minnow', 'common stickleback', 'cushy', 'doctor', 'eastern stickleback', 'European stickleback', 'jacksharp', 'New York stickleback', 'pinfish', 'prickley', 'prickly', 'prickly back', 'pinkeen', 'saw-finned stickleback', 'spanicle', 'spannistickle', 'spanny', 'spannytickle', 'spantickle', 'sparnicle', 'sparny', 'sparnytickle', 'spawn', 'spawnykettle', 'spawnytickle', 'sprickleback', 'stickling', 'stickle-back', 'thornback' and 'thorny back'.

By and large, these names do not differentiate three-spined and ten-spined sticklebacks in places where both species occur.

Elsewhere across Europe, the common names applied to the three-spined stickleback include:

French	Arselet, Cordonnier, Crève-valet, Épinart, Épinglet, Épinoche, Épinoche, Épinoche à trois épines, Estancelin, Estranglo cat, Spinaubé, Spinavaou, Stichling
Dutch	Driedoornige stekelbaars, Stekelbaars
German	Dreistacheliger Stichling, Dreistachliger Stichling, Großer Stichling, Rotzbarsch, Seestichling, Stachel de butz, Stachelbauch, Stachele, Stachelfisch, Stechbüttel, Stecherling, Stechert, Steckbedel, Steckbüdel, Steckelbars, Steckelstange, Steckker, Stichling
Danish	Trepigget hundestejle
Swedish	Storspigg

Once again, the German language is rich in terms for another little fish.

Three-spined sticklebacks and the arts

The three-spined stickleback features in a collage of other fishes painted by A.F. Lydon (Fig. 8.2), illustrating the Reverend W. Houghton's 1879 book *British Fresh-Water Fishes* but widely reproduced and used for a range of purposes ever since.

A painting by an anonymous artist of the three-spined stickleback is also included in the tea cards series released by the Brooke Bond company in 1960 (Fig. 8.3), and this too is widely appreciated and has been reused extensively ever since.

Three-spined sticklebacks as food

Though sometimes numerous, three-spined sticklebacks are not only diminutive in individual size but also extremely bony and adorned with robust 'thorns'

Fig. 8.2. Painting of a three-spined stickleback by the artist A.F. Lydon.

Fig. 8.3. Painting of a three-spined stickleback appearing in a set of 1960 Brooke Bond tea cards.

on the back and belly. Kenneth Mansfield wrote in his 1958 book *Small Fry and Bait Fish: How to Catch Them* of the gastronomic virtues of this fish, stating that,

> I know no one who has eaten a frying of sticklebacks nor have I read anything about their edibility. No doubt, if a man were forced to eat them by necessity he would find them sustaining, for they are oily fish.

Mansfield continues,

> In the Scandinavian countries and in Germany oil is extracted and fishmeal made from sticklebacks in areas where they are particularly abundant.

It seems unlikely that these practices continue nearly seven decades on from when this was written. I can find no evidence to suggest that they still do.

Three-spined sticklebacks as fishy fertilizer

In days gone by, where stickleback abounded in slow-running or still waters in the east of England, sticklebacks were harvested as a fertilizer for arable fields. In his 1808 book *The Complete Angler's Vade-Mecum*, Captain T. Williamson writes that sticklebacks were,

> sold by the bushel, as manure, both in Lincolnshire and in Cambridgeshire.

This is corroborated by Thomas Frederick Salter in his 1815 book *The Angler's Guide*, writing that,

> Pricklebacks are frequently used, in Lincolnshire, for manure, being always very numerous in the fens; but sometimes, they become so numerous as to make it necessary to separate and find new situations, which happens once in eight years, upon an average; during which migration, part of the river Welland is almost choked with them, at which time they are collected in nets, sieves, baskets, &c., to the amount of cart loads, and spread on the land as manure, and, I am informed, fertilize it extremely.

Though unfortunate for sticklebacks, their presence in such clear abundance means that harvesting gluts of this small fish served as an efficient, natural means to transfer nutrients as a fertilizer from watercourses on to adjacent cropped land.

Three-spined sticklebacks as pets

Three-spined sticklebacks make hardy and fascinating pets, adapting well to captivity and exhibiting their full range of behaviours in unheated aquaria. Neither will they outgrow their tanks, though it is essential that only one male is present, or that the tank is big enough to accommodate multiple territories, when the males take on their gaudy mating livery and become highly territorial. A potentially limiting factor though is their predatory behaviour, necessitating finding for them a constant source of live food such as mosquito larvae, water fleas or small worms or maggots.

Given how showy and hardy these little fishes are, and their fascinating behaviour, it is constantly surprising how fishkeepers in countries in which they occur overlook them in favour of species imported from overseas.

Three-spined sticklebacks and nature conservation

Three-spined sticklebacks are assessed on the IUCN Red List (The International Union for Conservation of Nature's Red List of Threatened Species) as of 'Least Concern' (LC) in terms of their risk of extinction, as they are common, widespread and there is no evidence of declining populations. Neither are they listed explicitly under any British or European nature conservation legislation.

Three-spined sticklebacks (Fig. 8.4) though are thought to be a pest in the Black Sea region of south-eastern Europe due to concerns about their consumption of the young of fish species caught for human consumption, and as they are also thought to compete for food with these other fish.

Fig. 8.4. Outline of a three-spined stickleback. (Image © Mark Everard.)

Ten-Spined Sticklebacks: The Shy Cousin

9

I have some sympathy with the ten-spined, or nine-spined, stickleback (Fig. 9.1). It is a smaller and shyer creature than its more brazen three-spined cousin, which hardly helps it being anything other than serially neglected. I have also to confess that, even in my own 2008 book, *The Little Book of Little Fishes* celebrating the often-neglected little fishes, I lumped Britain's two freshwater species of stickleback into the same chapter. In part, this was justified by the fact that their biology is so similar, avoiding unnecessary repetition. But another factor is that I have never lived in parts of the country where this retiring creature is abundant. These omissions collectively have hardly done much to raise the profile of this frequently disregarded little fish.

In fact, the ten-spined stickleback has numerous unique features. Not the least of these are disagreements in the common name about the number of spines it possesses but also, as we will see, its record size. The ten-spined stickleback is certainly worthy of more respect, despite its more retiring disposition and smaller size.

Male ten-spined sticklebacks also go through a similar sort of transformation, albeit less flamboyant than their three-spined cousins, taking on a pattern of bold black-and-white markings.

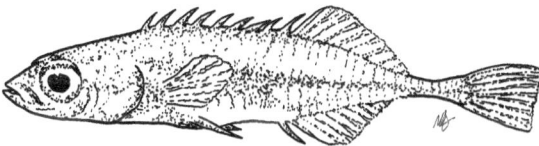

Fig. 9.1. The ten-spined stickleback. (Image © Mark Everard.)

© Mark Everard 2025. *Small Fry: Britain's Tiniest Freshwater Fishes* (M. Everard)
DOI: 10.1079/9781836991700.0009

Natural History of the Ten-Spined Stickleback

What is a ten-spined stickleback?

The ten-spined stickleback, also known as the 'nine-spined stickleback' or 'ninespine stickleback', goes by the Latin name *Pungitius pungitius* (Linnaeus, 1758). It is a member of the stickleback and tubesnout family (Gasterosteidae). Along with the three-spined stickleback, the ten-spined stickleback is one of two species from this family found in British fresh waters. A third native British stickleback, the fifteen-spined stickleback, occurs in coastal waters.

The head of the ten-spined stickleback is elongated leading to a small terminal mouth, and the eye is large. The front dorsal fin is modified into a row of between six and 12 spines that are separated from each other, each less stout and proportionately shorter than those of the three-spined stickleback. The posterior dorsal fin is supported by nine to 13 soft rays (0/9-13). Ten-spined sticklebacks generally have a regular dull or silvery body colour. The flanks of this fish lack scutes, though scutes are present on either side of the caudal peduncle (the narrow 'wrist' where the body meets the tail fin). The general body colour is muted, varying from pale green to grey or olive on the back and silvery below, though there may be irregular bars or blotches on the flanks. By contrast, the fins are colourless. Body colour can change during mating and when males become aggressive, as we shall see.

Ten-spined sticklebacks are small fish, with a maximum recorded length of 9 centimetres (3½ inches) though they are virtually always far smaller than this. This maximum length is, in fact, a record, making the ten-spined stickleback the smallest fully adult native British freshwater fish. (Sunbleak, addressed later in this book, also reach only 9 centimetres in length, but are discounted from the record book as they are an alien invasive species.)

Ten-spined stickleback distribution, habits and diet

Ten-spined sticklebacks have a wide circumarctic distribution in the northern hemisphere. This includes in Arctic and Atlantic drainages across North America including the Pacific coast of Alaska and the Great Lakes basin. Across Eurasia, ten-spined sticklebacks occur in coastal and lowland areas of northern Europe, including southern Norway and the Baltic basin, and as far eastwards as Russia, Siberia and Japan.

This small fish is retiring in behaviour and favours habitats such as small, weeded ponds, ditches and river backwaters holding permanent water, thriving in well-vegetated water bodies that may also be of low water quality. This nurturing environment enables them to evade predation as well as larger, open-water fish species with which they generally do not compete well.

Ten-spined sticklebacks are found throughout Britain, except northern Scotland. They can tolerate brackish water but, unlike their three-spined cousins, do not thrive in fully marine conditions so are naturally absent from the island of Ireland and have not thus far been introduced there by people.

Ten-spined sticklebacks are occasionally shoaling fish, though generally within the confines of well-vegetated habitats including in river edges, small streams, still waters of all sizes and lower-salinity regions of some estuaries.

Ten-spined stickleback reproduction and development

In many respects, the spawning habits of the ten-spined stickleback mirror those of the three-spined stickleback and – with apologies to the ten-spined stickleback – are therefore not repeated here other than to note some relatively minor differences.

During the spring and early summer breeding seasons, or when expressing aggression to other sticklebacks, male ten-spined sticklebacks darken, often with pronounced mottling or bars. They can though become totally black, or black on the belly, with the exception of the colourless fins. Female ten-spined sticklebacks retain their overall drab coloration in the breeding season.

Ten-spined sticklebacks are not long-lived fish, with a maximum recorded age of 5 years.

Ten-spined stickleback as prey and predator

The prey of the ten-spined stickleback and also its predators mirror those of the three-spined stickleback.

Ten-Spined Sticklebacks and Angling

Where they are present in any density, a good old-fashioned pond net on a long handle is about the best method for catching ten-spined sticklebacks.

To be honest, I know of only a few people, apart from me and Jack Perks (whose photographs grace this book), who seriously seek out ten-spined sticklebacks on rod and line. They are just so small, so getting them to grab hold of a bait is a real problem. A bigger problem is that, when three-spined sticklebacks are present, they tend to nab the bait first, being that much larger and more aggressive. The challenge is out there for the intrepid angler!

Ten-Spined Sticklebacks and Society

I regret that three-spined and ten-spined sticklebacks are barely distinguished by other authors outside the scientific community. Also, many of the facets of the ten-spined stickleback of interest to society are the same, or assumed to be the same, as those of the more common three-spined stickleback. Rather than

Fig. 9.2. Painting of a ten-spined stickleback by the artist A.F. Lydon.

repeat them, I will therefore indicate similarities and differences, with my apologies to the ten-spined stickleback for inadvertently perpetuating its perception as a poor relative of its more robust and widespread three-spined cousin!

Other names by which ten-spined sticklebacks are known

Virtually all of the common names noted for the three-spined stickleback across the English-speaking world are also applied to the ten-spined stickleback. Other common names by which ten-spined sticklebacks are known across a number of European countries, omitting those that duplicate those for the three-spined stickleback without mentioning the number of spines, include:

French	Épinoche à neuf épines, Épinochette, Kleine stichling, Marichaud, Petite épinoche
Dutch	Tiendoornige stekelbaars
German	Kleiner (Neunstacheliger) Stichling, Kleiner Stichling, Neunstacheliger Stichling, Seestichling, Steckling, Steekerling, Steekling, Steigbügel, Stichbeutel, Stichlinsky, Stickelbars
Danish	Nippigget hundestejle, Nipigget hundestejle
Swedish	Småspigg

Once again, the German language is rich in common names honouring this species and others comprising its wealth of little fishes.

Ten-spined sticklebacks and the arts

The ten-spined stickleback is included in a collage of other sticklebacks painted by A.F. Lydon (Fig. 9.2), illustrating the Reverend W. Houghton's 1879 book *British Fresh-Water Fishes*. A.F. Lydon's artwork has been reproduced and used widely ever since.

Fig. 9.3. Painting of a ten-spined stickleback appearing in a set of 1960 Brooke Bond tea cards.

The ten-spined stickleback also features as part of the set of fish illustrations by an unidentified artist on tea cards issued by the Brooke Bond company in 1960 (Fig. 9.3). These images have also been widely appreciated and reused ever since.

Ten-spined sticklebacks as pets

In common with their three-spined cousins, ten-spined sticklebacks, being small and hardy, adapt well to aquaria and can be fascinating pets though they do require small live food to keep them healthy. In captivity, as small fishes, they exhibit the full range of their wild behaviours including breeding and raising young. A note of caution is repeated here: you should ensure that the tank is large enough to account for the territorial behaviour of male fish, or else only to keep a single male in with a group of females.

Ten-spined sticklebacks and nature conservation

Under the IUCN Red List (The International Union for Conservation of Nature's Red List of Threatened Species), a comprehensive and regularly updated inventory of the global conservation status of species of plants and animals, ten-spined sticklebacks are assessed as of 'Least Concern' (LC) owing to stable populations found across their distribution range.

Ten-spined sticklebacks to not feature explicitly in other strands of nature conservation legislation.

Ten-spined sticklebacks as record-breakers

As noted when introducing this small and shy species, ten-spined sticklebacks attain a maximum recorded length of only 9 centimetres (3½ inches), making them the smallest fully adult native British freshwater fish (Fig. 9.4). As also noted, sunbleak, discussed later in this book, attain the same maximum length but are excluded from competition as they are an alien invasive species in British waters.

Fig. 9.4. Outline of a ten-spined stickleback. (Image © Mark Everard.)

Ruffe: Spiky Ruffians
10

The ruffe (Fig. 10.1), along other smaller fish species naturally occurring in British fresh waters, are often dismissed as 'minor species'. This rather misses the point that all species are important, and all are fascinating in their own right. Such is the case with the ruffe. Despite this, ruffe do not feature in many books on coarse fishes. They are, for example, absent from Arthur P. Bell's 1926 *Fresh-water Fishing for the Beginner*, Eric Marshall-Hardy's 1943 *Coarse Fish*, and Nick Giles' 1994 *Freshwater Fish of the British Isles: A Guide for Anglers and Naturalists*.

I sought to rectify this situation by their inclusion not only in some of my more general books on fish, for example my 2020 book *The Complex Lives of British Freshwater Fishes*, but also with a sole focus on ruffe in my 2023 book *Ruffe: The Spiky Freshwater Ruffian*. In my ruffe book, I posed the question "Does a little freshwater fish like a ruffe deserve a book all of its own?" While the fish may not command a massive market attractive to large publishers, I concluded strongly that, based on the merits of this quirky little fish, it most definitely did deserve a book dedicated to its cause and that I would make sure that this happened!

Ruffe may be small fish, drab even in coloration and sometimes regarded as a nuisance by anglers seeking bigger fish, but they have character for sure

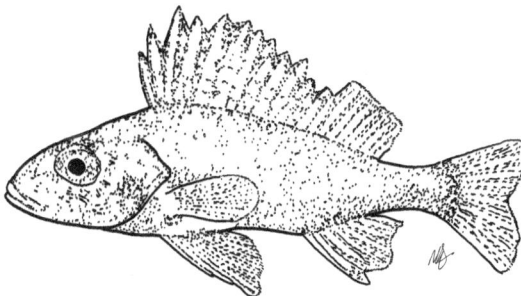

Fig. 10.1. The ruffe. (Image © Mark Everard.)

© Mark Everard 2025. *Small Fry: Britain's Tiniest Freshwater Fishes* (M. Everard)
DOI: 10.1079/9781836991700.0010

as well as some challenging aspects for nature conservation. They are indeed 'spiky little freshwater ruffians', whether loved by anglers and naturalists or loathed for threats posed when introduced beyond their native range.

Natural History of the Ruffe

What is a ruffe?

The ruffe known to British anglers and naturalists is most properly referred to as the 'Eurasian ruffe', going by the Latin name *Gymnocephalus cernua* (Linnaeus, 1758). However, across Europe and Asia, there are five species within the genus *Gymnocephalus*, all of them variously known as ruffe. The Eurasian ruffe is the only ruffe species that occurs in British waters and is the species that will preoccupy us in this chapter. So, where you see the word 'ruffe' without other qualification, this is the familiar species we are talking about.

Ruffe are members of the perch family (Percidae). They are not large fish, with a maximum recorded length of 25 centimetres (almost 10 inches), but mainly they are far smaller in British waters. For reference, the British record rod-caught ruffe at the time of writing (winter 2024/2025) weighed 149 grams (5 ounces 4 drams), caught from West View Lake in Cumbria in 1980.

Like all members of the perch family, ruffe are primarily a freshwater fish though with some tolerance of conditions in lower-salinity estuaries. The body shape of the ruffe is approximately oval in cross section and elongated, tapering from the front half but with a high back on which two dorsal fins are located. The flanks are fully covered by small scales and are slightly rough to the touch. The body colour of the ruffe varies considerably depending on the environment in which it lives. In clear water conditions, body colour can range from olive-brown to golden-brown or khaki on the back and flanks, overlain with numerous small and irregular dark blotches, darker on the back and fading to lighter tones on the belly. By contrast, in turbid canals and other murky waters, camouflage colour is less critical due to low light penetration, so the general colour can fade to a largely silvery or grey background overlain with darker blotches. A continuous lateral line extends the whole length of the body on either flank, along which are 33–42 lateral-line scales. In addition to the lateral-line sensory organs, Alwyne Wheeler notes the presence of additional sensory canals in his 1969 book *The Fishes of the British Isles and North-West Europe*,

> A series of large sensory canals just beneath the skin in the head are particularly noticeable below the eye and on the pre-operculum.

Like all members of the perch family, European ruffe have two dorsal fins. The front dorsal fin comprises 11–19 spines and the rear dorsal fin is supported by 11–16 soft rays. These two dorsal fins are fused together and connected without a notch (cumulatively XI-XIX/11-16 in scientific notation). This is a differentiating feature from small perch, which have a clear gap between the front and rear dorsal fins. On the underside and to the rear, the anal fin of the ruffe is supported by two spines and five or six soft rays (II/5-6), while the caudal (tail) fin has 16 or 17 soft rays. The ruffe generally also has dark spots on the membranes between the spines and soft rays of the dorsal fins, mirroring the body coloration which continues into the tail fin. The ruffe's other fins – pectoral, ventral and anal – tend to be paler in colour.

The head of the ruffe is rounded and lacking scales, bearing the same pigmentation as the body, and is interspersed with conspicuous slime-filled sensory canals beneath the skin that can at times make this fish slimy or 'snotty' to the touch (as we will review later in this chapter). The mouth is small and positioned in a slightly inferior (down-turned) orientation, lacking barbels (sensory 'whiskers'). The jaw is armed with a row of teeth, even though tiny, and vomerine teeth (also known as palatine teeth) are present as a small patch on the roof of the mouth.

The eye of the ruffe is distinctive, both relatively large and also glassy in appearance. Also, as Kenneth Mansfield records in his 1958 book *Small Fry and Bait Fish: How to Catch Them*,

> An unusual feature of the colour scheme of this fish is the purple eye.

Kenneth Mansfield continues, noting that,

> They appear to find their food more by sight than smell and are daylight feeders.

Potential confusion with the perch (and gudgeon)

There are similarities in form between the 239 members of the perch family found across the northern hemisphere. In particular in British waters, confusion can occur between ruffe and small perch (*Perca fluviatilis*), juveniles of which in murky waters often lack their normal striking coloration including distinctive strong vertical stripes on the flanks. With familiarity, the differences become obvious, but numerous authors note potential mistaken identity. Izaak Walton, for example, wrote in his classic 1653 book *The Compleat Angler*,

> There is also another Fish called a Pope, and by some a Ruffe, a fish that is not known to be in some Rivers; it is much like the Pearch for his shape, and taken to be better than the Pearch, but it will not grow to be bigger than a Gudgion; he is an excellent Fish.

As noted above, the conjoined dorsal fins of the ruffe are a key distinguishing feature. Another is that the rear of the ruffe's preoperculum (the front bony plate above the main bony gill cover, or operculum) is scalloped and spinous, described by Kenneth Mansfield in his 1958 book *Small Fry and Bait Fish: How to Catch Them*,

> The plates of the forward gill covers are scalloped, each point forming a short spine. The rear gill cover tapers back into a single larger spine.

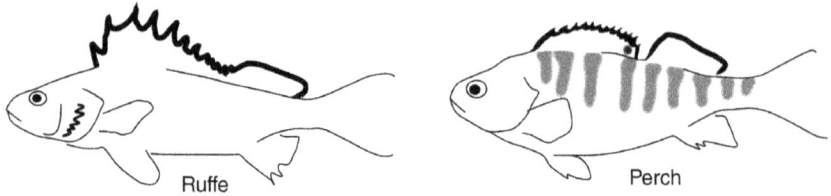

Fig. 10.2. Key distinguishing features of ruffe (left) and perch (right). (Image © Mark Everard.)

The number of spines on the preoperculum also serves as a feature differentiating the five ruffe species found across northern Europe and Asia.

A simple illustration highlighting these differences – the continuum between dorsal fins and the scalloped preoperculum, and the bold stripes generally found on the perch – is presented in Fig. 10.2 'borrowed' from my book *Ruffe: The Spiky Freshwater Ruffian*.

In his 1879 book *British Fresh-Water Fishes*, the Reverend W. Houghton notes a former mistaken belief by some that ruffe were hybrids between perch and gudgeon, possessing an overlap of the features of both. This former misplaced belief, or reference to superficial appearance, is reflected in an alternative French common name 'Goujon-perche' (or 'Gudgeon-perch'). Houghton wrote that,

> The Ruffe, resembling in outward characters and markings both the Perch and the Gudgeon, has sometimes, but erroneously, been considered a hybrid between the two; although there is actual proof that closely allied species do, occasionally and perhaps not infrequently, cross, this is not true of fishes so distantly related as the Perch and the Gudgeon.

Ruffe distribution, habits, diet and senses

The (Eurasian) ruffe occurs naturally throughout Europe in the Caspian, Black, Baltic and North Sea basins, including in Great Britain where it is the only native ruffe species, and as far northwards as approximately 69°N in Scandinavia. It is also found in Asia in the Aral, Caspian and Black Sea basins including the rivers, lakes and brackish coastal waters of most of the former USSR, and in the Arctic Ocean basin eastward to the Kolyma drainage.

Through introductions, both intentional and accidental, Eurasian ruffe are also now found in several other countries including parts of western France that were not naturally colonized, in northern Italy and Greece as well as in North America. Ruffe, like many other animal and plant species, can be problematic when introduced and established beyond their native range into ecosystems within which they have not co-evolved. The specific impacts of introduced ruffe are reviewed later in this chapter when considering their implications for nature conservation.

The natural range of ruffe in Britain reflects the geological history of our islands, and in particular the former connection with what are now continental European river catchments via the 'Doggerland Bridge' inundated at the end of the last ice age (between 6500 and 6200 BC).

This same geological history explains the natural absence of ruffe from the island of Ireland, fortunately with no known introductions at the time of writing. The geographic separation of Ireland accounts for the fish listed as present in Ireland by the medieval clergyman and chronicler Giraldus Cambrensis in his *The History and Topography of Ireland*, with the ruffe an absentee as it lacks fully sea-going life stages and hence was unable to cross the Irish Sea.

Ruffe then are principally fishes of fresh water, found in a wide variety of river, lake, pond and canal habitats. They also occur in the brackish waters of some estuaries up to around one-third the salinity of seawater and are common in lower-salinity zones of the Baltic Sea region. Ruffe are bottom-living fish, more commonly found in enriched waters with a soft bottom and also withstanding a limited degree of pollution. For this reason, ruffe often prosper in the turbid waters of canals. Though far from rare, ruffe distribution is patchy, despite they may be locally common and sometimes very abundant. Ruffe tend to be social, forming dense shoals in suitable waters. Illustrating this tendency, Richard Franck recorded in his 1694 *Northern Memoirs* that,

> Ruff for the most part move all in a body. One would think them mutineers, because all of a piece; for if you hang but one, all the rest are in danger. Nor will they revolt, or retreat from their diet, since every one resolves to eat till he die.

Ruffe feed exclusively on a variety of small animals, comprising many types of invertebrates including larval and adult insects, crustaceans, worms, and opportunistically on fish eggs and fry as well as the tadpoles of newts, frogs and toads. This tendency to feed on the eggs and fry of fish can cause problems, as we will see later in this chapter. Though equipped with a lateral line and additional sensory canals in the head to detect vibrations in the water, ruffe have keen sight and tend to be active by day relying largely upon sight to hunt prey. By night, they are generally less active, evading detection by their predators, though may still hunt by sensing vibrations.

Ruffe reproduction and development

Ruffe display little or no sexual differentiation, though females tend to be slightly larger than males as well as living a little longer. In the spring and early summer, generally in a period between March and the end of May depending on latitude and water temperature, mature ruffe head into marginal water to spawn. They spawn communally at depths less than 3 metres (about 10 feet).

Spawning occurs in spring, female ruffe releasing strands of eggs that are fertilized on release by male fish. These strips of eggs, each egg tending to be white or yellow in colour and ranging from 0.34 to 1.3 millimetres (around

0.01 to 0.05 inches) in diameter, become sticky on contact with water, allowing them to adhere to hard submerged surfaces such as stones, plants, dead branches, underwater roots and tree branches dipping into the water. Female ruffe may lay their eggs in two or more phases during the summer, separated typically by a period of around 30 days. Ruffe are highly prolific fish but exhibit no parental care once the eggs are released, and so there is consequently heavy predation by invertebrates and fish. As noted by Kenneth Mansfield in his 1958 book *Small Fry and Bait Fish: How to Catch Them*,

> For so small a fish the ruffe produces an inordinate number of eggs – 150,000 to 220,000 have been mentioned – but many are consumed by fish and insect larvae.

Those eggs that survive hatch after five to 30 days, depending on water temperature. The tiny and incompletely developed hatchlings remain immobile for between three and seven days consuming their attached yolk sac. Then, typically a week after hatching, the larvae swim up into open water to feed on planktonic crustaceans and other small invertebrates. After an initial few days' feeding, growing and developing in open water, the juveniles then metamorphose and move to the bed. These metamorphosed juveniles are initially solitary, feeding on progressively larger animal prey.

As they further develop, the solitary habit is replaced with social, shoaling behaviour. Ruffe mature after two or three years, potentially doing so after only a year of life in warmer waters. Ruffe are not long-lived, typically reaching up to 5 to 6 years old, though the maximum recorded age is 10 years. Female ruffe tend to be longer-lived than male fish.

Ruffe predators

Like many smaller species of fish, ruffe can fall prey to a range of predatory and opportunistic omnivorous fishes, their spiky front dorsal fin and gill covers affording some dissuasion though these fish are small enough to be swallowed whole by larger predators. Additional predators include piscivorous birds ranging from kingfishers to egrets, herons, mergansers, goosanders and cormorants. Otters are among the mammals potentially preying on ruffe, the benthic habit of this fish making it vulnerable to the 'truffling' feeding habit of the otter on the beds and margins of water bodies.

Ruffe and Angling

While ruffe may lack the muscle power of a barbel or the delicate manners of a roach, they are none the less generally willing to take an offered worm, maggot or other small live animal bait even in quite chilly conditions. A ruffe may be particularly welcome to a match angler when tough conditions dissuade other fish from biting, whether as part of a mixed bag or specifically targeted. At other times though, ruffe can be a nuisance when present in profusion, their voracious habits intercepting baits intended for larger fish. As Alwyne Wheeler put it in his 1969 book *The Fishes of the British Isles and North-West Europe,*

> The ruffe is too small to possess any interest to anglers; indeed its bold biting at bait intended for better fish is more often a nuisance than anything.

Locating ruffe

As ruffe favour still or sluggish water and tend to congregate around features such as steep banks and fallen wood in the water, these are the places to seek them out. On canals, look for edges and drop-offs rather than open water. Above all, make sure the bait is on the bottom as this is not a fish that will intercept a bait any appreciable distance above it.

Ruffe baits

As ruffe are dedicated carnivores, maggots and worms are handy and readily sourced baits when fishing for them. The colour of the maggots seems not to matter a great deal, though I prefer red ones as these are closest in appearance to the bloodworms (larvae of chironomid midges) that ruffe prey upon in muddy beds. Equally, almost any type of worm, or sections of larger worms, are accepted. However, I have found that brandlings seem to be particularly attractive despite, or perhaps because of, the pungent yellow liquid they can exude. Izaak Walton's Piscator in the 1653 *The Compleat Angler* would seem to concur, noting

> You must fish for him with a small red worm; and if you bait the ground with earth, it is excellent.

Other small invertebrate baits can also be deployed if you are so minded. These can include freshwater shrimps, caddis larvae and bloodworms, presented on fine hooks to avoid killing the bait.

Ruffe tackle and tactics

Akin to virtually all of the species addressed in this book, no special tactics or tackle are required when fishing for ruffe. A simple float outfit, fished on a running line or from a pole or whip, can present a bait in the margins of the water you are fishing. Either that, or a light leger using sensitive bite indication such as a quivertip. A few loose offerings of the bait you are presenting on the hook, or ground bait laced with fragments of these baits, may attract ruffe and get them seeking out the titbit on your hook. But avoid too big a hook as the mouth of the ruffe is surprisingly small for such a voracious little fish. Also avoid excessive ground bait as this may draw in other fish species that you may not be targeting.

In reality, ruffe are far from fussy eaters, and nor are they hard to catch if present in any numbers. In his 1958 book *Small Fry and Bait Fish: How to Catch Them*, Kenneth Mansfield recommends,

> "The lightest of tackle can be used, with a small float. The bait should fish on or very close to the bottom. It is advisable to strike as soon as the float bobs in order to lip-hook the fish: if a ruffe is given a few seconds spare it will gorge the bait.

> "Most of the ruffe caught are taken by anglers seeking more important quarry and they can undoubtedly be a nuisance, especially in waters where they subsist in thousands, as in some parts of the Broads."

I think it is important to reiterate the advice about a rapid strike, not because the fish bites rapidly but as the bait can otherwise be swallowed deeply – distressing for both ruffe and angler alike! And, while on the subject, I will once again repeat a sound bit of advice given to me many years ago for fishing in general and that I have lived by since:

> Always carry three disgorgers: one to use, one as a spare, and a third to give to any other angler you meet who does not have one!

As predators actively hunting by sight, ruffe can also be caught on small soft plastic lures as well as artificial flies. These should be fished close to the bed and near the kinds of features that these fish prefer, such as vertical canal banks and around woody debris. My preference is the smallest soft plastic lures of up to 2½ centimetres (1 inch) long mounted on small, weighed jig heads and 'walked' across the bed mimicking a bloodworm. Alternatively, drop-shotting tactics can be used to the same effect. Some anglers far more expert than me at this approach swear by dipping their lures in flavourings to increase attraction. This is specialist angling for 'mini-monsters', and for more details I cannot do better than refer you to the highly informative 2019 book *Hooked on Lure Fishing* by Dominic Garnett and Andy Mytton.

Ruffe used as bait

Ruffe have reputedly been used as live baits when fishing for predatory fish, particularly pike. In fact, pike anglers are frequently blamed for introducing ruffe into the English Lake District, Loch Lomond in Scotland and Llyn Tegid (Bala Lake) in Wales. This is unproven, and I am among several other angling authors casting doubt on this assumption. While ruffe are easily sourced where they are abundant, and are hardy and so easy to transport, they are in fact small when compared to baits normally used by pike anglers. Of this possibly misplaced assumption about pike anglers translocating ruffe as live baits, Kenneth Mansfield wrote in his 1958 book *Small Fry and Bait Fish: How to Catch Them* that,

> As far as I know ruffe are never used for live-bait except when nothing else is available.

An undated article by the Canal and River Trust endorses this view,

> Whilst I would not pretend to have much pike fishing knowledge, I have yet to come across a single pike angler who has ever claimed to have used ruffe as a livebait.

The fact remains that ruffe can theoretically be used as live baits, or indeed dead baits, though they are not a preferred option, lacking the attractive silvery sheen, oiliness and larger bulk of many other fish used for this purpose.

Ruffe and Society

Ruffe are fascinating little fishes from biological and piscatorial points of view. However, they have far wider meanings and cultural associations. Here, we explore some of these broader associations between ruffe and people.

Other names by which ruffe are known

Throughout history, as with many species, a succession of scientific Latin names has been applied to ruffe, all now superseded by *G. cernua*. The Latin name of the genus, *Gymnocephalus*, comprises 'gymno' (meaning 'naked', 'bare' or 'exposed', derived from the Greek) and 'cephalus' (meaning 'head'), reflecting the thin, scaleless covering of skin on the head of the various species of ruffe.

The common name 'ruffe' is believed to have derived from a former spelling of 'rough', the scales of this fish being notably rough to the touch.

However, the ruffe is also blessed with a range of common local names in Britain. Across Britain, ruffe are also widely known as 'pope', the origins of which are uncertain, but 'pope' is an affectionate term meaning 'father' and may perhaps be related to another common alternative name 'daddy ruffe'. Other local British nicknames include 'ruff', 'jack ruffe' and 'Tommy ruffe'. Another not uncommon alternative name is 'snotties', celebrating the sometimes rich mucous exuded by this little fish.

Ruffe also go by a range of common names in various languages across their wide geographical range, including:

French	Frash, Goujon-perche, Grémeuille
Dutch	Pos
German	Gries, Hork, Kaulbarsch, Kauschbarsch, Kugelbarsch, Kuhlbarsch, Kulbersch, Kulberschke, Kutt, Kütteberschi, Pfaffenlaus, Pfifferling, Posch, Rauiegel, Rotzbarb, Rotzbarsch, Rötzert, Rotzkater, Rotzwolf, Schlickerbarsch, Schnotterbarsch, Schnotterboars, Schroll, Steuerbarsch, Stuhrt, Stune, Stur, Sturbarsch, Tork
Danish	Hork, Almindelig hork
Swedish	Gärs

Once again, the German language excels in the diversity of common names it bestows on this little fish!

Ruffe and the arts

I have searched high and low for masterwork paintings of ruffe and found none, though some old woodcuts of varying quality could qualify. Equally, I have sought symphonies, concerti, lieder or other musical repertoires dedicated to the ruffe, and also drawn a blank.

However, ruffe are included in a collage of larger fishes painted by A.F. Lydon (Fig. 10.3) which illustrated the Reverend W. Houghton's 1879 book *British Fresh-Water Fishes* and has been reproduced and used widely ever since.

Fig. 10.3. Painting of a ruffe by the artist A.F. Lydon.

Fig. 10.4. Painting of a ruffe appearing in a set of 1960 Brooke Bond tea cards.

The ruffe also features among a set of tea cards issued by the Brooke Bond company in 1960 (Fig. 10.4), these paintings being widely appreciated and reused ever since.

In terms of books dedicated to the ruffe, there seems to be only my own 2023 *Ruffe: The Spiky Little Freshwater Ruffian*. However, far more celebrated writers have made reference to ruffe. One such is the Russian short-story writer, playwright and physician Anton Pavlovich Chekhov (1860–1904), famed in particular for his short stories. Chekhov was in fact a keen angler, and his various writings have referenced gudgeon, chub, perch, carp (*Cyprinus carpio*), burbot (*Lota lota*) as well as ruffe. Extracts from two of Chekhov's early stories mentioning the humble ruffe are reproduced here:

> Lord, what sport that is! To catch a burbot or chub, say, is like coming across your own long-lost brother! And every fish has its own mentality, you know: one of them you catch with a live-bait, another with a grub, another with a frog or a grasshopper. You've got to know all about that! Take the burbot, for example. The burbot's not a choosy fish, she'll go for a ruffe even, whilst the pike is fond of a gudgeon and the asp a butterfly.

> Here is Pschanka, I remember, you used to catch pike a good two foot long, and there were burbot, ide and bream, all decent-size fish too, but now you're grateful if you catch a jack-pike or a perch six inches long. You don't see a proper ruffe even.

Ruffe as food

There is a long history of exploitation of ruffe as human food, particularly across central, north and west Europe and into Russia, though these prac-tices have since largely disappeared from use along with their associated traditions.

For all its perhaps less-than-delicious external appearance, many are the writers throughout history who have extolled the virtues of ruffe as human food. The earliest would appear to be Dame Juliana Berners, noting of the ruffe in her 1496 *Treatyse of Fysshynge with an Angle*,

> The Ruf is a right and holsom fysshe.

A century and a half later, Izaak Walton wrote in his 1653 *The Compleat Angler* that,

> he is an excellent Fish; no Fish that swims is of a pleasanter taste.

In 1958, Kenneth Mansfield writes in *Small Fry and Bait Fish: How to Catch Them* that,

> In north Germany the larger ruffe from brackish waters are considered a luxury and form a local dish of some repute.

Alwyne Wheeler recorded in his 1969 book *The Fishes of the British Isles and North-West Europe* that,

> Ruffe are said to be good eating.

There is, then, consensus that ruffe are good eating, though I have to admit that I have not tried eating them. Various authors commend their firm tasty flesh, especially when fried, with others recommending ruffe soup. For example, J.H. Keene documented in his 1881 book *The Practical Fisherman* that,

> I much prefer the ruffe, so far as its flavour is concerned, to even the delicate sweet gudgeon or its cousin, the perch. It, however, requires careful cooking, not that it is a fish to which it is necessary to add all sorts of condiments, but because to over-fry it or over-bake it (with bay and rosemary) is to spoil a certain nutty flavour which a ruffe from the clear river at the latter end of July possesses.

To this, an essay 'A Kettle of Fish' from the 1929 compilation *Rod and Line* by the author Arthur Ransome noted that,

> In Russia to this day a kettle of fish (ukhá) is often the most important part of a banquet. Curiously enough the two fish chiefly prized for this purpose are two which are not often eaten in England, the ruffe and the burbot.

Kenneth Mansfield further endorsed the pleasing flavour of ruffe, suggesting a recipe in his 1958 book *Small Fry and Bait Fishes: How to Catch Them*,

> the angler who finds himself plagued with ruffe could do worse than set to and catch a couple of dozen for his supper. The fish, while still complete, should be washed free of mucus and allowed to dry. Then cut off the heads and clean the insides using as little water as possible. Dry them, and fry in hot, shallow fat, oil or butter. When served, the skin, with its tough scales and spined fins, can easily be removed. On this account it is better not to use batter or egg and breadcrumbs in their preparation.

Bent J. Muus and Preben Dahlstrom wrote in their 1967 *Collins Guide to the Freshwater Fishes of Britain and Europe* that,

> Its flesh is tasty, especially when fried, but the fish is too small to be valuable as a table fish, although ruffe soup can be recommended.

For those hankering to know more, a comprehensive resource concerning the use of ruffe as human food throughout history can be found in a scientific paper published by Svanberg and Locker in 2020, titled 'Caviar, soup and other dishes made of Eurasian ruffe, *Gymnocephalus cernua* (Linnaeus, 1758): forgotten foodstuff in central, north and west Europe and its possible revival' (this paper is fully referenced in the Bibliography at the end of this book). Among examples of recipes and culinary uses of ruffe, Svanberg and Locker (2020) also record that,

> One can still buy dried ruffe chips in Estonia. Dried fish are a traditional beer snack in Russia and the Baltic States. At Tallinn Airport small plastic bags with salt-dried ruffe were sold in the spring of 2019.

Commercial harvesting of ruffe

Given the long history of exploitation of ruffe, it is not surprising to learn that there was formerly an associated history of commercial harvesting. Ruffe were formerly exploited widely for food in North Germany, with further important fisheries in eastern Europe. In the mid-twentieth century, commercial fishing for ruffe was carried out in numerous fisheries in Russia including in Kursian Bay (Baltic Sea basin) and numerous lakes and reservoirs in the Volga, Ob and Yenisei catchments.

Bent J. Muus and Preben Dahlstrom relate an interesting aspect of ruffe fishing in their 1967 *Collins Guide to the Freshwater Fishes of Britain and Europe*,

> At one time the fishery in eastern Europe was important. In the eastern Prussian haffs the ruffe which seem sensitive to sounds, were formerly driven into stake-traps by 'clap boards' (the banging of boards which have ends sticking into the bottom).

However, Alwyne Wheeler records of ruffe fisheries in his 1969 book *The Fishes of the British Isles and North-West Europe* that,

> It is not commercially fished, except that at one time in North Germany it was exploited for food.

There may now be no remaining commercial ruffe fisheries, with the majority of interest in fisheries circles relating to ruffe as a potential competitor for food with other more economically valuable fish species, or as a predator of their fry and eggs.

Ruffe and nature conservation

Under the IUCN Red List (The International Union for Conservation of Nature's Red List of Threatened Species), a comprehensive and regularly updated inventory of extinction risk at global scale of plants and animals, ruffe are assessed as of 'Least Concern' (LC). They are, like all other small species, none the less important components of functional ecosystems, with both positive and in some places negative implications, though more often they are completely overlooked.

Ruffe are not listed under the European Union (EU) Bern (or Berne) Convention (on the Conservation of European Wildlife and Natural Habitats). However, the scarcer schraetzer or 'striped ruffe' (*Gymnocephalus schraetzer*) and the Danube ruffe (*Gymnocephalus baloni*), both species found in continental Europe but not in the British Isles, feature in the schedules of the Bern Convention. Neither are European ruffe scheduled under the EU Habitats Directive (Council Directive 92/43/EEC on the Conservation of Natural Habitats and of Wild Fauna and Flora), though the schraetzer is listed in two Annexes as an indicator of habitats of concern.

If they are not a priority species for protection, European ruffe have certainly attracted attention in terms of the threats they pose to other fish species. This is related to their propensity to become rapidly established when introduced to new regions, and their habit of feeding on the eggs and fry of other fish as well as competing with these species for food resources.

Problems arise from the spread of ruffe to places in which they are not native, and the pressures they place on the species and the balance of ecosystems found there. While ruffe are part of the British fish fauna, they are not naturally part of aquatic ecosystem in northern England. They have though become established in the English Lake District for a matter of decades. Introductions of ruffe into the deep, cool lakes found there has been hugely detrimental to native species of fish adapted to these post-glacial habitats – in particular European whitefish (*Coregonus lavaretus*), vendace (*Coregonus albula*) and Arctic charr (*Salvelinus alpinus*) – that come to the lake margins and tributary streams to spawn but are now confronted by egg and fry predation by introduced coarse fish species including ruffe and roach. The vendace is now confirmed as extinct from its last English stronghold in Bassenthwaite Lake, with the introduced ruffe a primary culprit along with increasing agricultural pollution. The same pressures are impinging on the 'gwyniad' (the local name for the European whitefish) in Llyn Tegid (Bala Lake) in Wales. Similar concerns are raised in Scotland where ruffe, first discovered in Loch Lomond in 1982, prey mainly on invertebrates on the loch bed as well as the juveniles and eggs of native fish species, significantly including the 'powan' (the local name of the European whitefish). Elsewhere across England, ruffe are still spreading westwards from their original distribution in eastern-flowing catchments.

Although ruffe are naturally widespread across mainland Europe, they are an introduced species in Lake Geneva (straddling Switzerland and France), Lake Constance (the Bodensee bordering Austria, Germany and Switzerland), Lake Mildevatn (Norway), the Camargue region of southern France, and northern Italy. They are raising concerns in these places for similar reasons.

Of particular concern is the introduction and rapid spread of ruffe into North America. As these fish are hardy, without exacting water-quality requirements, grow rapidly and become sexually mature at an early age, they have attributes making them prone to invasion. There is a suspicion that ruffe may have found their way into the American Great Lakes in the ballast water of ships. Ruffe were first reported in Lake Superior in 1987, and these fish have since spread in that lake as well as being found in Lake Huron and Lake

Michigan bordering the US states of Minnesota, Wisconsin and Michigan and the Canadian province of Ontario. Significant research has been devoted to the spread of ruffe, seeking to eradicate them from the Great Lakes network due to concerns about their potential for damage to native fish species through competition for food and habitat and also predation upon native sportfish eggs with associated socio-economic ramifications. Ruffe have already become the most numerous fish species in the St. Louis River watershed at the head of the Great Lakes. Chemical controls using both poison and pheromones are being trialled, as is increasing the population of native predatory fish.

Ruffe as pets

Ruffe are one of those native fishes that adapt well to, and do not rapidly outgrow, a home aquarium. Surprisingly, they are somewhat hesitant feeders – quite distinct from the habits they display in the wild – and can be easily muscled out of food by more voracious native fish such as small perch, so care should be taken to ensure they receive their fair share of food. They are dedicated carnivores, so items such as small worms, maggots, mosquito larvae and pieces of frozen prawn will all be accepted. Dr Günther Sterba in his 1962 book *Freshwater Fishes of the World*, writing initially about perch, recommended

> midge and other insect-larvae, worms and slugs; larger individuals will also take young fishes

then continued later in the book to write that 'Ruff' [*sic*] and 'Schraetzer' (a ruffe species from mainland Europe),

> can also be kept successfully for a long time under the conditions just described.

Ruffe social organizations

At the time of writing, there are two virtual digital communities (Facebook social media pages) of people interested in the ruffe, indicating social cohesion around ruffe-related interests. However, unlike pike, roach, carp, tench, perch, barbel and gudgeon, there is to date no society dedicated solely to the ruffe.

A ruffe story: 'The Great Ruffe Hunt'

I have struggled somewhat to find a published story specifically about ruffe, as distinct from stories such as those of Chekhov and other writers that refer to ruffe only tangentially.

The dedicated ruffe story that I did find was titled 'The Great Ruffe Hunt'. This tale is to be found in my own out-of-print 2008 book *The Little Book of Little Fishes*. The story relates, literally, to a 'great ruffe hunt' in a Westcountry canal involving me, my good angling buddy Sid and my then 4½-year-old daughter. This story is reproduced here for you to enjoy!

"We had been in a state of intermittent yet mounting anticipation for some months now, building up to this momentous specimen-hunting trip. Now, with early spring threatening to crash our grand Good Friday party, we were on our way at last.

"Me, Sid, Daisy and a stack of fishing tackle, in the car, singing songs. Well, to be more accurate, Daisy was singing the songs. A mixture of kiddies TV ditties, the snippets of Christmas carols that she'd learned at school, and a fine rendition of 'Wild Thing' that she'd picked up from hearing my band rehearse and play. I had always viewed it as a key part of my parental responsibility to school her in the finest of culture.

"In no time, we had found the lay-by and parked up overlooking the steep-sided valley, in the bottom of which lay the mist-veiled mystery of the fishery we had been told of in hushed tones.

"Well, sort of.

"It was misty, that much was true, though the sun was already beginning to burn through the early spring haze with the promise of a glorious day ahead. However, it would not be entirely honest to maintain that the tones informing me of this venue were hushed. Indeed, a mate of mine had shouted it from his 4×4's window over the noise of the two-litre engine. The mystery of the fishery is also, in truth, a little overstated. It was, after all, only a canal, a murky one at that and already buzzing with narrow boats, its towpath a motorway of dog walkers and cyclists. In fact, the biggest mystery was how to avoid the dog crap on the towpath. But hell, why ruin a good story with trivial detail?

"Two and more months prior, back in the clutches of a cold January evening, I had mentioned to a roach-fishing mate of mine that I had a book to write on small fry. However, I'd continued, I had not caught a Tommy Ruffe in the fifteen years since I had last fished the tidal Thames. Where, I wondered, could a Westcountry boy like me find a ruffe so that I could photograph and draw it? Mike was the right guy to ask, having recently retired as a Fisheries Officer; he knew the rivers, streams, lakes, pools and canals of north Wessex like the back of his hand. And, it came to light, he knew a thing or two about local ruffe.

"Now, before I go on here, I should perhaps point out that, naturally, the ruffe is a resident of the slacker waters to the east and south of England, although it has recently spread into more northerly waters. Where it has appeared out of place, it has sometimes wrought dreadful environmental harm.

"Take the vendace, for example. The vendace is an exceedingly rare whitefish occurring in just one lake in England, Bassenthwaite, in the English Lake District where, since the introduction and subsequent population explosion of ruffe from the 1980s onwards, its survival is severely jeopardised. It is not the poor old ruffe's fault of course. Not merely because predation on the eggs and fry of vendace is but one of a catalogue of environmental problems threatening the very existence of the vendace, albeit one of the most important contributory factors. More to the point, all the ruffe are doing is surviving, as nature has equipped them to do. It is man that is, yet again, the guilty party for upsetting delicate ecological balances evolved through millennia with profound and unpredictable results. Taking the ruffe to new places was a wanton, albeit probably innocent in intent, act of environmental vandalism that is likely to rob future generations of a unique place, ecosystem and species."

(Sadly, this part of my 2008 story about the impacts of ruffe introduction to Bassenthwaite was salutary: vendace have since been declared extinct from that English lake.)

"So the ruffe is not natural to the Westcountry, and not a species that I could normally expect to encounter without a long-distance specialist campaign which the size of the fish – up to just a few ounces – hardly justifies. Having said that, I do feel a little ashamed about the 'size-ist' attitude that that statement belies! But in reality it is a little like making a two-hundred mile round trip for sticklebacks. On reflection, since I have, on various occasions, made a similar investment of time and fuel to completely blank out on distant waters, maybe a long-range stint after ever-obliging sticklebacks is not such a daft idea after all! However, back to the ruffe ...

"So, Mike said, there were ruffe here in this busy Westcountry canal, courtesy of another inadvertent introduction. The canal's owners and operators, it seems, had decided to restock the canal a couple of years previously. Unsurprisingly, they took the cost-effective route of transferring excess fish from other of the many canals and balancing ponds they owned across the country, with appropriate health checks and blessings from the relevant environmental authorities. However, it maybe speaks volumes for the rigour of the checks that the trans-ferred silver fish also enjoyed the undetected company of a little spiky intruder. The little spiky intruder had found the new waters, relatively pure for a canal, very much to its liking and – or so rumour had it – had prospered. They were, I was told, present and doing rather well, though they might take a little diligence to locate.

"A challenge then, but one made more attainable by its proximity to our home village. With Sid, my neighbour and sometime fishing buddy, we had plotted this day over several months around my kitchen table with Daisy. And, after Daisy's bed-time, or else when faced with a particularly profound question of angling technique, Sid and I had often found it necessary to continue our deliberations in the village pub!

"Now, we were hell-bent on ruffe, no matter what the difficulties. We were the team to display the necessary grit, staying power, fine-tuned hunting instinct, resilience and repertoire of playground songs to bring an elusive ruffe to net. We were people with a purpose. This was to be The Great Ruffe Hunt. With capital letters!

"We finished the song we were singing ('Wild Thing' for the N-th time) and got out of the car, donning clothing appropriate for a Great Ruffe Hunt, shouldering the tackle and bait, and picking up the sundry other essentials. These 'sundry other essentials' naturally included a huge bag of sweets, snacks and drinks, at least as big as all the piles of tackle put together, to assuage the constant grazing needs of a four-and-a-half year old girl.

"What Daisy lacked in age and height she certainly always made up for in energy, enthusiasm and volume. And anyway, even accounting for Daisy's tender years, the combined age of The Great Ruffe Hunt team was one hundred and thirteen. Between us, we felt we were not lacking in the depth of experience and commitment necessary to winkle out the most obdurate of ruffe!

"We dropped down the steep path to the turning basin, where the canal flexed through ninety degrees and the spur to another long-abandoned canal had once branched off to the north Somerset coalfields. This long-defunct arm of water had offered the deep coal of Somerset a smooth, slow passage through to the Thames. This was, however, for just a few short years before the railway rendered it no more than a footnote in history, albeit an important one if we consider the consequences of its excavation for the development of modern geology.

"Normally, I eschew busy waters in favour of quiet retreats, hidden depths and unknown specimens. So, as you may imagine, it was quite an eye-opener to join the thronging flow of boats, dogs, bicycles and walkers, already quite a bustle at little past a civilised breakfast time. But here we were, doing as needs must for The Great Ruffe Hunt!

"Where to start? My eyes are, though I say it myself, reasonably well attuned for watercraft on moving waters, spotting weed, ripple and eddy for clues as to what lies beneath and the places a fish may rest or feed. Also, practice and biological knowledge help me somewhat on still waters, locating likely interception points from patches of plant growth, water and bird movement, topping fish and other, often subliminal clues. But this was a wholly different order of challenge! The brown waters, like cold, flavourless hot chocolate, were of uniform appearance wherever I cared to gaze. Everywhere had, evidently, the same width and bank structure, and probably the same depth. Not a shoot of submerged vegetation was visible, nor indeed was it even likely to be as the whole turbid mass was regularly stirred to its very depths by the sporadic passage of boats. Watercraft then...

"We walked upstream, for no other reason than that there were fewer boats going that way. This took our merry band across the magnificent aqueduct that carried the canal high about the Bristol Avon, with fabulous views in either direction. It was, indeed, an engineering wonder. But we were, we felt, now seeking at least as imposing an angling wonder to locate and to prise ruffe from this strange water!

"Shortly, we came to a tight turn in the canal where it made another ninety-degree bend to resume its easterly course on the opposite side of the gorge framing the river valley below. It was broader here, slightly reedy in the margins, and dead, wind-blown vegetation gathered in small islets on its surface. More promising! Immediately upstream of the wide elbow of the canal – where 'upstream' is of course just a term to describe the direction in which we were walking – we found a long-abandoned lock, its timbers decayed in situ and the canal a little constricted. A feature!! We dropped the angling kit gladly, pausing for a breather and to wipe away the sweat of an unexpectedly warm early spring day as the sun now broke through the fading morning haze. This seemed as likely a spot for a first assault on the water, though it would doubtless be a mere pause on the protracted pursuit that was to be The Great Ruffe Hunt.

"Daisy, meanwhile, had kept up a perpetual running commentary on everything we had seen and passed by, observations tumbling over questions and exclamations with no pause to heed any of the answers she had, apparently, urgently sought not seconds before. All in all, our spirits were high. And this despite the scale of the challenge facing us to find the gold at the end of the rainbow that was The Great Ruffe Hunt.

"The spot we had chosen had the added advantage of being tree-lined on both sides. A line of tall beeches stood on the outside of the towpath to our backs, lining the edge before the scarp fell away steeply downwards again to the river in the gorge below. On the opposite side of the canal, the bank ascended majestically from the water's edge, upwards through a beech forest carpeted with copper-gold fallen leaves from the season gone. The horizon vanished high above us in a mist of interlacing, steel-grey beech boughs, naked to their bud and bark in the first stirrings of spring. We were secure in a cocoon of tall, leafless trees to each side, in a slightly quieter corner of the busy Bank Holiday canal side. Who could ask for more?

"Well, we could of course! We were, after all, embarked on The Great Ruffe Hunt, and were now poised and ready to pit our wits with this small alien from the Eastern counties. As we assembled tackle, a proud cock pheasant announced his presence nearby with its strident barking crow. He strutted the far bank of the canal, an iridescent adornment to the burnished backdrop, not fifteen yards away yet not so much oblivious to us as fearless in the security afforded by the narrow barrier of water. Not even a passing narrow boat, engine sputtering and bilge pump spattering, turned its head. Daisy waved and said hello to both the pheasant and the narrow boat. Only the narrow boat's occupants felt inclined to wave back, or even to bother to notice us.

"I loosened the terminal tackle from its fixing on Daisy's telescopic six-foot rod, giving line from the bright blue reel of which she was so proud, and she was the first of our intrepid team of hunters to wet a maggot in earnest in The Great Ruffe Hunt. She was the first to turn anticipation and stealth, watercraft and guile, into wet line and persistence. And surely persistence was to be our watchword if The Great Ruffe Hunt was to bear fruit. We were here for the duration: resilient, inventive, enquiring, eventually to outwit the quarry that we had been discussing and plotting to intercept for months now!

"I turned to loosen the rod ties on my own waggler rod, pushing together the carbon sections and baiting the hook.

"And then Daisy's float sailed away.

"She lifted the rod tip and connected to ... a Tommy Ruffe!

"The fish struggled vainly on the end of Daisy's line as she swung it towards me to unhook. She had done it, and within seconds, before Sid and I, the seasoned campaigners, had even wet a line!

"Daisy, a tenth of my age, had whooped my ass big time! For a moment, the eager anticipation of The Great Ruffe Hunt was deflated, but any sense of disappointment was instantly dissipated by my pride at my little girl's achievement. We had won! We had succeeded within the first few seconds of The Great Ruffe Hunt!

"I studied the fish closely, photographing it with Daisy holding it, and on its own, giving it a little water and then making a rough sketch. I took in its frilly gill edge, green mottling, fused pair of dorsal fins held proudly aloft, and the pair of big, dark eyes.

"'Good to see you again, old friend!' I thought to myself. 'It's been a long time!'

"Then, I popped it into the net. This, I just knew, was going to be a good day!

"Without labouring the detail, it was indeed a good day which saw us land literally hundreds of fish between us, from, amazingly, bleak and dace, through to perch, lots more ruffe, bream and roach, and gudgeon galore. The hour or so that Sid and I thought we might get from Daisy stretched out to nearly six hours before I, in an instant of piscatorial laxity through which parental duty was able to shine through momentarily, decided we really should reel in and get this little girl fed properly.

"We talked about our day all the way home. We are talking about it now, weeks later, and I am sure will talk about it for many years to come. We had shared a little long moment of Westcountry magic, hunting by intent small fry that we might otherwise have walked by or simply dismissed as not worth the hassle. Yet they had given us back so much in reward for a little simple appreciation.

"And so ended The Great Ruffe Hunt, a continuing reminder of how great small fish can be, of the pleasures that might unfold unexpectedly, like vivid rose petals from a tight bud, on discovering a new and hidden face of a venue that one might more often overlook. So, the sun set at last on a fantastic day of good company and good sport.

"Thank you, small fish everywhere.

"Thank you, The Great Ruffe Hunt!"

Appreciation for the spiky ruffian

The ruffe (Fig. 10.5) may be a fish of small stature, loved by some and loathed by others, but is a fish with many fascinating attributes and roles to play in ecosystems.

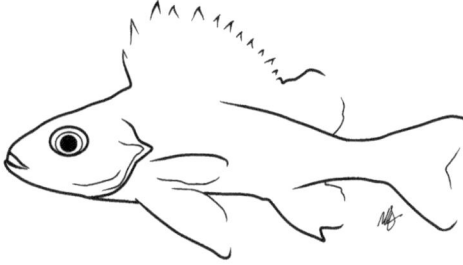

Fig. 10.5. Outline of a ruffe. (Image © Mark Everard.)

Brook Lamprey: Small Mysteries in the River Mud

11

Are lampreys (Fig. 11.1) even fish?

It may surprise many people that, after a lifetime working on the topic, the American evolutionary biologist Stephen Jay Gould concluded, "There is no such thing as a fish"!

Though we may feel we know what a fish is, the reality is that the term 'fish' is applied to around 32,000 species of limbless vertebrates found in aquatic habitats around the world. Many are from a bewildering diversity of evolutionary lines. As the requirements of living in a watery world impose similar pressure in terms of mobility and ability to breathe, there is quite a bit of convergence around common functional features.

Gills, or at least gill-like structures (some fishes can also breathe air), are required for the exchange of gases with surrounding water. Fins, or at least fin-like appendages, are needed for propulsion in a more viscous medium, which is also aided by a streamlined body. So many evolutionary lines have converged in the development of 'fish-like' features over the past 500 million years. However, from a genetic point of view, a salmon is a far closer relative to a camel than it is to a hagfish, though both are familiar to us as 'fishes'.

This point is directly relevant to the brook lamprey, as indeed to all the other lamprey species found around the world. Whereas many true 'fish' are teleosts (bony fishes), the lampreys not only lack bones but also lack jaws, reflecting a far more distant evolutionary lineage. Add to this the cryptic habits of lamprey larvae, known as ammocoetes, largely living hidden in river margins and silty depths, and the brook lamprey becomes a mysterious beast indeed!

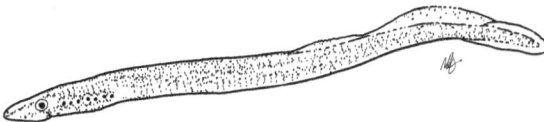

Fig. 11.1. The brook lamprey. (Image © Mark Everard.)

© Mark Everard 2025. *Small Fry: Britain's Tiniest Freshwater Fishes* (M. Everard)
DOI: 10.1079/9781836991700.0011

Over the years, I have had the pleasure of many encounters with these cryptic and enigmatic little fishes. I have found the curious ammocoete larvae when I have disturbed sediment in the river margin, either incidentally or when sampling river invertebrates. The first time I saw a small gathering of metallic metamorphosed adult brook lampreys was many years ago when, stumbling across this charming vignette that put all thought of fishing to one side, I sat and watched a party of four brook lampreys making repeated and eventually successful attempts to ascend a cascade of water over the bedrock of the River Irfon above Builth Wells in Wales.

The fact remains that this is an elusive fish, often remaining a hidden secret in moving waters. Their secretive habits may mean that they are more widespread than commonly thought, secreted as small mysteries unsuspected in the river mud.

Natural History of the Brook Lamprey

What is a brook lamprey?

The brook lamprey, or more correctly the European brook lamprey, is known to science by the Latin name *Lampetra planeri* (Bloch, 1784). It is one of three species of lamprey from the lamprey family (Petromyzontidae) found in British fresh waters. All three of these lamprey species spawn in fresh waters and have similar freshwater larval stages but two of them – the sea lamprey (*Petromyzon marinus*) and the river lamprey (*Lampetra fluviatilis*) – migrate to sea after metamorphosing, only returning to rivers to spawn.

The bodies of adult lamprey species are eel-like, lacking scales. The lampreys also lack paired fins and have a skeleton that is cartilaginous rather than bone. The lampreys all lack jaws, instead having a circular disc. There is also only as a single nostril, located on the top of the head. Curious primaeval fish indeed, if in fact the term 'fish' accurately applies!

Adult brook lampreys are small fish with a maximum length of 20 centimetres (about 8 inches) and are found in rivers throughout the British Isles except the far north of Scotland. In their adult form, brook lampreys are slightly metallic in colour, with a pair of functional eyes and a small sucking disc with blunt, weak teeth. There are seven small, round gill openings lacking gill covers on either side of the head behind the eye. These round gill openings were once thought to be additional eyes, hence the alternative name 'nine eyes'.

However, this is a fish that spends virtually all its life in a larval stage known as an ammocoete. Brook lamprey ammocoetes grow a little longer than the adult form. They are jawless, toothless and blind with little pigmentation, and they possess two dorsal fins the rearmost of which merges into the tail fin. Brook lamprey ammocoetes also have the line of seven circular gill openings behind each eye seen in the fish's adult life stage (Fig. 11.2). Ammocoetes live buried in river silt, filter-feeding on fine suspended matter in the passing water

Fig. 11.2. An ammocoete larva. (Image © Jack Perks.)

or grazing on detritus. After living for five or six years as larvae, brook lampreys metamorphose into the eel-shaped adult form in spring.

Unlike the other two British lamprey species, brook lampreys do not migrate after metamorphosing. Neither do they feed as adults, instead spawning communally in river gravels after metamorphosis, generally after a short upstream migration, and dying shortly thereafter.

Brook lamprey distribution, habits, diet and senses

Brook lampreys are to be found in well-oxygenated flowing water, typically in the middle and upper reaches of rivers and small streams but also occasionally in lakes. The ammocoete larvae are strongly associated with detritus-rich sandy or clay sediments in which they burrow, and upon which they feed in addition to filter-feeding from the flowing water above.

Brook lampreys are found across European rivers draining into the North Sea as far north as Scotland and mid-way up the Norwegian coast, as well as in the Baltic Sea basin and in Atlantic drainages as far south as Portugal. There are also populations in French and Italian river basins draining into the Mediterranean.

Many lamprey species have marine phases. Despite the brook lamprey having no migratory phase and dying shortly after metamorphosing to spawn, they are naturally present in Ireland despite the lack of a land or river bridge with mainland Britain by reason of the 'satellite species' phenomenon addressed later in this chapter.

The pair of eyes of the brook lamprey are small in the adult form. However, as a filter-feeding and surface-scraping feeder during its prolonged larval phase, most of the senses generally relied on are chemical and tactile. Indeed, another common name applied to this fish is 'blind lamprey'.

Brook lamprey reproduction and development

Reproduction is a fatal affair for all species of lamprey. This reflects an early evolutionary stage of development as these ancient fishes have shown little change when compared with fossils dating back over 360 million years, way back in the Carboniferous Period. This is true of the two British species with sea-going adult forms, as well as the brook lamprey which lacks an extended adult life stage.

All British lamprey species spawn in the gravels of cool, well-oxygenated streams or larger clean rivers. Typically, this takes place in April or May, brook lampreys metamorphosing into the adult form shortly in advance and generally migrating upstream prior to spawning. Once a suitable gravel-bedded spawning substrate is found, male and female brook lampreys collaborate in the building of spawning pits, known as redds, grasping rocks one by one and moving them clear of the redd. They may also use their tails to fan away smaller particles. When the redd is dug, the adult lampreys twine their bodies together, brook lampreys often doing so communally, releasing eggs and sperm simultaneously. After the eggs are released and fertilized, many fall into the redd. Their tasks completed, the adult brook lampreys – as is the case for all lamprey species – then die shortly thereafter.

After between one and two weeks, ammocoete larvae emerge from the eggs and seek out soft silt into which they burrow. Once secure in this silt habitat, the ammocoetes feed by filtering microorganisms, algae and detritus from the water and the soft silt sediment.

One of the many fascinating aspects of the biology of non-migratory and migratory species of lamprey in most genera is that, among the 38 extant species of lamprey distributed around the world, there are pairs of species: one parasitic and migratory and the other non-parasitic and permanently resident in fresh water. The ammocoete larvae of these 'satellite species', also known 'paired species', are to all intents and purposes identical both morphologically and also genetically. Although they inhabit the same spawning and larval habitats, these differing lamprey life strategies may be to some extent isolated from each other as discrete breeding populations. This view is supported by the fact that lampreys have been found to choose mates of a similar body size. Further details about lamprey satellites need to be resolved, though current opinion is that these pairs of satellite species may in fact be the same species with two distinct life strategies. (See Box 11.1 for more technical detail.)

> **Box 11.1.** The distinctiveness of lamprey 'satellite' species
>
> The larval stages of the European brook lamprey (*Lampetra planeri*) and river lamprey (*Lampetra fluviatilis*) cannot be distinguished morphologically, and various studies have also found that they are virtually indistinguishable genetically (Espanhol *et al.*, 2007). Similar conclusions have been drawn from the genetic profiles of satellite lamprey species from Japan, North America and Australia (Docker *et al.*, 1999; Hwang *et al.*, 2013; Carim *et al.*, 2023; Carpenter-Bundhoo and Moffatt, 2024).
>
> While some studies identify minor variances in genetic markers between European brook lamprey and river lamprey also noting hybridization between the two 'species' (Mateus *et al.*, 2013; Souissi *et al.*, 2022), analysis of genetic differences between European brook lamprey and river lamprey suggests that these lampreys are part of a 'species complex' (a group of closely related organisms so similar in appearance and other features that the boundaries between them are often unclear) comprising distinct ecotypes in which different phenotypical features are triggered by environmental factors (Torresen *et al.*, 2025). Minor genetic variations observed between European brook lamprey and river lamprey sampled across their broad European range appear to be more significant than variations between paired species in single localities (Pereira *et al.*, 2011; De Cahsan *et al.*, 2020), supporting the species complex hypothesis.
>
> The references cited can be found in full in the Bibliography at the end of this book.

Brook lamprey as prey and predator

'Predator' may be too grandiose a term to apply to the primarily filter-feeding and sediment-scraping habits of brook lamprey ammocoetes. However, small and microscopic animals – 'water fleas' and other tiny crustaceans, rotifers, tardigrades and others – form part of this soup of material ingested by the larvae.

Both ammocoetes and metamorphosed brook lampreys though have their opportunistic predators, including fish, birds and mammals. These predatory fish include pike. I have observed this, much to my surprise, one day when quite a decent-sized pike hit my spinning lure on the Hampshire Avon. The pike was unexpectedly thin and, when I banked it to remove my lure, the reason became clear. Another angler's poorly tied trace with two treble hooks was lodged in the throat of this unfortunate fish. Of course, I removed the offending hooks – a lucky day for that ravenous pike though I am not quite sure it appreciated it at the time! – but I had to first clear out what initially looked like a few softened willow leaves to gain access to the hooks. I was though amazed to see that these were no leaves, but were a half-dozen ammocoete larvae all of them around 7.5 to 10 centimetres (3 or 4 inches) long. The pike, all green and yellow mottled fury, was duly returned with minimal time out of the water to live and, importantly, be able to feed another day and so to have benefitted massively from the encounter. Then, as I sat back down on the moist leaf mould, while a passing family of long-tailed tits twittered among the hazel twigs overhead, I spent a little time looking closer at the curious little larvae that had been trapped by the hooks in the pike's throat. All were of course dead by that time, but indicative both of a healthy population in the river, and that they were actively hunted by hungry pike regardless of their cryptic habits and small size. I admired the row of seven small, round gill openings on each side of the head behind the little eye, the small mouth disc, the eel-like body, the continuous fins from back to belly, and the slightly metallic colour reminiscent of a slow worm. A small mystery in the river mud, revealed to me only due to an unusual set of circumstances.

On another river, I have also had perch as well as eels regurgitate ammocoetes when caught on rod and line. Ammocoetes then are fair game for many a predatory fish. I am sure that egrets would not refuse one of these larvae as they stalk stream margins looking for the subtle movements of small fish and larger invertebrates. Perhaps otters too with their 'truffling' feeding behaviour, though I have not seen direct evidence of this.

Brook Lampreys and Angling

The term 'angling' does not really apply when seeking out brook lampreys as they lack jaws and true mouths, the larvae largely filter-feeding and scraping detritus from the silt and the adults lacking a functional mouth. The word 'angling' is derived from fishing with an 'angle', or hook. Therefore, there is no feasible way of angling for brook lampreys.

None the less, where present, brook lampreys and the ammocoete larvae of other lamprey species turn up not infrequently in nets used for surveying river invertebrates when 'kick sampling'. Most often, this is in the silty river margins in which they burrow and feed. If you are keen on seeing an ammocoete, these silty margins are the places to look. If you want to observe or catch a metamorphosed brook lamprey, by hand or net, then the sandy or gravelled streams and river edges into which they spawn in springtime are the places to go.

Various historic texts note that brook lampreys can be used as bait. However, given the legal protections that this fish now enjoys this is no longer acceptable. Collecting sufficient for a day's fishing is, anyhow, unlikely.

Brook Lampreys and Society

Lamprey species generically have an interesting relationship with humanity. They have a wide range of affectionate and other local names, depictions in art, as well as a reported history in gastronomy and in aquaculture. Their history as pet fish is, in reality, probably based on a misinterpretation of a name. The lampreys are also of nature conservation interest.

Other names by which brook lampreys are known

Brook lampreys go by various local names. One of them is 'Planer's lamprey', named after the German zoologist Johann Christian Planer who first recognized the species (originally as *Petromyzon planeri*) in 1784.

The name 'nine eyes' is curious as, if one adds each of the small 'real' eyes to the seven round gill openings behind it on each side, one obtains a total of eight rather than nine eyes. If the single round nostril on the top of the head is counted, this could be nine viewed from each side of the fish. The exact mathematics and biology probably evaded those so nicknaming the fish!

Somewhat more accurately (if we ignore the real eyes and just count the round gill openings on either side), the term 'seven eyes' is also used, for example as written by Thomas Frederick Salter in his 1815 book *The Angler's Guide: Being a Complete Practical Treatise on Angling* referring generically about lampreys,

> The lamprey eel is of the shape of the Lamprey or Seven Eyes ... The lamprey eel is frequently caught in the river Severn, near Gloucester...

Other of the many common names applied to this fish include 'European brook lamprey', 'mud lamprey', 'blind lamprey, 'pride', 'prid' and 'sandpride', and more simply and commonly 'brookie'.

Brook lampreys also have a range of common names in various languages across their wide geographical range, including:

French	Lamproie de ruisseau d'Europe, Lamproie de ruisseau européene, Petite lamproie
Dutch	Beekprik
German	Bachneunauge, Bachpricke, Kleines Neunauge, Neunauge, Neunäugler, Neunhocker, Queder
Danish	Almindelig bæklampret, Bæklampret, Bæknioje
Swedish	Bäcknejonöga

Brook lampreys and the arts

No major composer seems to have created a musical masterwork inspired by the humble brook lamprey, nor has any feted artist devoted their brushwork to this little fish. However, the Irish-born British watercolour artist Alexander Francis Lydon (1836/1837–1917) produced classic artworks of British freshwater fishes. These famously illustrate the Reverend W. Houghton's 1879 book *British Fresh-Water Fishes*, and have been widely used for many purposes including, for example, by the Portmeirion Pottery company for their *Compleat Angler* range of tableware, though the image of the brook lamprey (Fig. 11.3) was perhaps deemed insufficiently appealing for its appearance on the dinner table.

Fig. 11.3. Painting of a brook lamprey by the artist A.F. Lydon.

Fig. 11.4. Painting of a brook lamprey appearing in a set of 1960 Brooke Bond tea cards.

A painting of the brook lamprey by an unnamed artist also features as part of the series of tea cards issued by the Brooke Bond company in 1960 (Fig. 11.4). These charming cards have been widely appreciated and reproduced ever since.

Brook lampreys as food and in aquaculture

Brook lampreys are reputed to have mediocre-quality meat and, of course, are of small size. Other, larger lamprey species have though a greater gastronomic pedigree.

King Henry I (1068/1069–1135), the fourth and youngest son of William the Conqueror, was King of England from 1100 until his death, which was reportedly hastened by eating a number of lampreys against his physician's advice.

King John (1166–1216) was reported to be particularly fond of lamprey pie. John was King of England from 1199 until his death shortly before his 50th birthday, attributed by various accounts to 'a surfeit of peaches and cider'.

However, the little brook lamprey is unlikely to have played any part in these unfortunate Royal departures.

Larger lamprey species are considered a delicacy in some parts of Europe, including south-western France, though brook lampreys are probably safe from such human predation. In passing, it is interesting to note that eels are not considered Kosher under Jewish law as they lack readily visible and detachable scales as well as true fins. The same should apply equally to lampreys. The status of these fishes is less clear under Islamic law, as the Halal (allowed) code specifies that the fish must come out of the water alive. However, as larger lamprey species subsist off the living flesh and body fluids of other fishes, this carnivory means that these fishes are theoretically Haram (forbidden).

Another interesting lamprey-related connection is that of Lucius Licinius Murena, Consul in the late Roman Republic in 151 BC. Pliny reportedly attributed the invention of fishponds to Murena. In fact, 'Murena' is a cognomen – an extra personal name given to an ancient Roman citizen – derived from the Latin word *murenae*, meaning a lamprey or an eel. However, given the unlikely reality of lampreys featuring in aquaculture, the 'murenae' in this case were more likely eels rather than lampreys. This does cast some doubt on whether the various food or other attributes assigned to lampreys might, in reality, relate to eels.

Brook lampreys as pets

I think it is fair to say that keeping brook lampreys as pets – as ammocoetes rather than as adults as these fish die shortly after metamorphosis and spawning – is something very much for the specialist. In fact, I know of no one who has kept such a pet, and this is anyhow not really acceptable given the nature conservation protections now afforded to this little fish. Ammocoetes also mainly lie buried in soft sediment, filter-feeding or scraping detritus from the adjacent surface, so are hardly likely to be a great spectacle!

An interesting entry in Izaak Walton's 1653 *The Compleat Angler* records that Sir Francis Bacon,

> in his History of Life and Death, mentions a Lamprey, belonging to the Roman emperor, to be made tame, and so kept for almost threescore years; and that such useful and pleasant observations were made of this Lamprey, that Crassus the orator, who kept her, lamented her death; and we read in Doctor Hakewill, that Hortensius was seen to weep at the death of a Lamprey that he had kept long, and loved exceedingly.

This type of intelligent behaviour, let alone the longevity of the specimen and the feasibility of keeping and feeding a parasitic species as a pet, certainly casts doubt on whether this animal was a lamprey. More likely, this is a mistranslation from 'murenae', which most likely and feasibly referred to an eel. Eels will certainly adopt such intelligent behaviours in captivity or indeed when routinely fed in the wild. One of my friends told me that a lady neighbour used to bang on a metal post in the river margin to call to an eel that had learned this was a signal that she had some food for it. Eels are therefore by far the more likely culprits.

Brook lampreys and nature conservation

Unfortunately, the only times that the protection of fishes seems to be taken seriously is in the cases of the few threatened species that are scheduled for protection under nature conservation legislation. Britain's three species of lamprey fall under various legally protective measures, as do two species of shad (family Clupeidae) and various members of the salmon family (family Salmonidae including two whitefish species).

Given its widespread distribution and lack of evidence of any obvious decline in population, the IUCN Red List (The International Union for Conservation of Nature's Red List of Threatened Species) assesses the brook lamprey as of 'Least Concern' (LC) in terms of extinction risk.

However, the European Union (EU) Bern Convention (on the Conservation of European Wildlife and Natural Habitats 1979) schedules all three species of lamprey that occur in British and European waters (the brook lamprey, the river lamprey and the sea lamprey) in Appendix III concerned with "species that are regulated but the exploitation of which is controlled in accordance with the Directive". All three lamprey species are also scheduled under the EU Habitats Directive (Council Directive 92/43/EEC on the Conservation of Natural Habitats and of Wild Fauna and Flora) under Annex II: 'Animal and plant species of community interest whose conservation requires the designation of Special Areas of Conservation (SAC)'.

The continuously mysterious brook lamprey

The brook lamprey (Fig. 11.5), and the lamprey fishes in general, are fishes of an ancient lineage, and perhaps not even fish at all if we were capable of defining exactly what constitutes a 'fish'. Cryptic and curious, the brook lamprey remains a mystery in the rivers and streams of Britain.

Fig. 11.5. Outline of a brook lamprey. (Image © Mark Everard.)

Smelt

Sand-smelt

Salty Visitors

12

Aside from the fully freshwater species listed in this book, two 'small fry' species visit estuaries and, sometimes, the lower reaches of British rivers particularly in the summer months.

In fact, estuaries (Fig. 12.1) are important recruitment and nursery areas for many marine fishes, and lower reaches of rivers closer to the sea are frequently exploited by some marine species capable of adapting to freshwater conditions.

Fig. 12.1. Estuarine habitat at low tide. (Image © Mark Everard.)

© Mark Everard 2025. *Small Fry: Britain's Tiniest Freshwater Fishes* (M. Everard)
DOI: 10.1079/9781836991700.0012

Significant among these are the flounder (*Platichthys flesus*), European seabass (*Dicentrarchus labrax*) and three species of mullet: the thick-lipped grey mullet (*Chelon labrosus*), the thin-lipped grey mullet (*Liza ramada*) and the golden-grey mullet (*Liza aurata*), all of which can penetrate some way into fully fresh water. These fishes are opportunists, feeding on the rich resources of summer estuaries but generally vacating them in the colder months when freshwater flows are also most often at their strongest and coolest. All of these summer visitors are also too large to qualify as 'small fry' species.

It is two smaller 'small fry' species that are of interest to us in this book. These are two entirely unrelated British smelt species with a strong association with estuaries, particularly when it comes to breeding.

The Smelt

The smelt (*Osmerus eperlanus* (Linnaeus, 1758)) (Fig. 12.2) enters the brackish water of estuaries, and occasionally lower rivers, to spawn between February and April. While adult smelt mainly return to sea after having spawned, the eggs they have deposited hatch into juvenile smelt that remain in estuaries during their early life stages. As they develop, they progressively move down towards the sea.

The smelt, also known as the 'sparling' or 'European smelt', is a small fish, weighing up to 250 grams (almost 9 ounces). It is the only member of the smelt family (Osmeridae) recorded in British waters. Smelt may be resident in some larger estuaries, also entering the lower reaches of rivers. They have a single dorsal fin lacking spines but supported by soft rays. The head and snout are pointed. The lower jaw has small teeth, the hind part reaching back to the hind margin of eye and the front of the jaw projecting forward a little beyond the tip of the upper toothed jaw. There is also an incomplete lateral line, and a pronounced silvery stripe runs along the bright flanks. Another of the smelt's characteristic features is that living and freshly caught fish are said to smell of

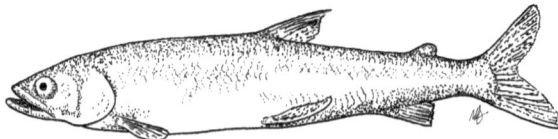

Fig. 12.2. The smelt. (Image © Mark Everard.)

cucumber. In addition to the usual complement of fins, smelt also have an adipose (fatty) fin between the dorsal and the tail.

Although the smelt is listed in the Reverend W. Houghton's 1879 *British Fresh-Water Fishes* as a species of fresh waters and some freshwater smelt populations occur across Europe, British smelt are inherently marine. They enter estuaries and occasionally lower rivers to spawn between February and April, adult fish then returning to the sea with juveniles also progressively moving down to estuaries after hatching.

The medieval clergyman and chronicler Giraldus Cambrensis, in his 1187 account of *The History and Topography of Ireland*, lists smelt as then present in Ireland, which is not surprising given they are principally a marine and estuarine species.

The smelt features in a collage of other fishes painted by Alexander Francis Lydon (Fig. 12.3), illustrating the Reverend W. Houghton's 1879 book *British Fresh-Water Fishes* and since widely appreciated and reused.

Smelt also feature within the set of tea cards issued by the Brooke Bond company in 1960 (Fig. 12.4). The images on these tea cards are widely appreciated and have been reproduced in many settings ever since.

Fig. 12.3. Painting of a smelt by the artist A.F. Lydon.

Fig. 12.4. Painting of a smelt appearing in a set of 1960 Brooke Bond tea cards.

Fig. 12.5. Outline of a smelt. (Image © Mark Everard.)

Smelt, though, have some reputation as a fine fish for eating. Kenneth Mansfield wrote in his 1958 book *Small Fry and Bait Fish: How to Catch Them* that,

> They are fish of very fine flavour which diminishes in proportion to the time they are kept.

Smelt are also a popular bait with pike anglers, available frozen in many tackle shops. They have a distinctive odour said to resemble that of cucumber, and their tough skin also ensures that dead baits do not fly off the hook too readily on casting or after prolonged immersion (Fig. 12.5).

The Sand-smelt

Despite a similarity of common name, the sand-smelt (*Atherina presbyter* Cuvier, 1829) (Fig. 12.6) is an entirely different species to the smelt. Also known as the 'little sand smelt', the sand-smelt belongs in a different family of fishes: the silversides family (Atherinidae).

Sand-smelt are a small (up to 100 grams or 3½ ounces) shoaling species of inshore and coastal marine waters, favouring estuaries and penetrating the lower reaches of rivers. The body of the sand-smelt has an intense silvery line, often outlined in black, from head to tail along a silvery, elongated and laterally compressed body that is covered in relatively large scales. The body lacks a true lateral line.

These fishes have two widely separated dorsal fins, the first with flexible spines and the second with one spine followed by soft rays, and a large pair of pectoral fins. The eyes are large, their diameter equal to the length of the snout, and the mouth is upturned. Sand-smelt are gregarious, favouring low-salinity water. They feed primarily on zooplankton, particularly small crustaceans and fish larvae. Isolated freshwater populations are known in several Italian and Spanish lakes and in some lower river reaches around the Mediterranean, though not in British waters.

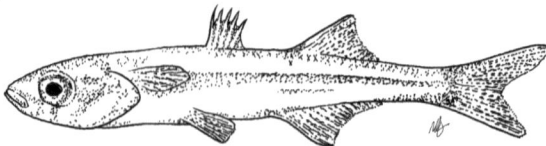

Fig. 12.6. The sand-smelt. (Image © Mark Everard.)

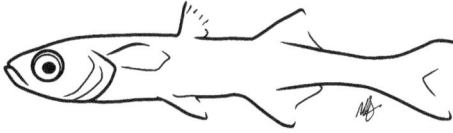

Fig. 12.7. Outline of a sand-smelt. (Image © Mark Everard.)

Sand-smelt (Fig. 12.7) reproduce in spring and summer, from May to July, favouring coastal lagoons or large intertidal pools, some of which may be in the lower reaches of estuaries.

Other Salt-Tolerant Fishes

As noted previously in this book, some other species can tolerate brackish conditions and may be found in upper estuaries. However, three-spined sticklebacks are tolerant of both fresh and saline waters. In fact, along coastlines with steep and short rivers that spate and are inhospitable during the winter months, three-spined sticklebacks may be flushed out into coastal waters to overwinter – a familiar sight off the west coast of Scotland.

Topmouth gudgeon

Sunbleak

Pumpkinseed

Bitterling

Brown goldfish

Gibel

Black bullhead

Small Invaders

In addition to our native fauna, a number of species of fish have been introduced into British fresh waters. Many of these, such as common carp (*Cyprinus carpio*), ide (*Leuciscus idellus*), wels catfish (*Silurus glanis*), rainbow trout (*Onchorhynchus mykiss*) and zander (*Sander luciperca*), are larger species. However, some smaller fish species are also now present and established. This chapter considers these small invaders in a little less detail than the mainstream of other British freshwater 'small fry' and also discusses the importance of controlling further invasions.

Longer-Term Visitors

Throughout history, people relocating to new regions of the world have brought with them species that are not native to these new environments. Some of these introductions have been deliberate, ranging from crop plants to livestock, including fish. One such is the common carp, the introduction of which into British waters is ascribed to monks in around the fourteenth century, principally for food, though various domesticated strains of carp are now spread for recreational angling and ornamental purposes. Various crayfish species have been deliberately introduced into British waters, particularly the American signal crayfish (*Pacifastacus leniusculus*) introduced to Europe in the 1960s to supplement crayfish fisheries, but which has subsequently caused serious damage to aquatic ecosystems.

Other species have hitch-hiked with people incidentally. These range from many ruderal weeds, beetles and other insects through to the European brown rat and American mink, as well as 'killer shrimps' and some fishes. Accidental fish introductions have often arrived via the aquarium and pond fish trade, as well as by recreational angling activities.

There is a significant potential for alien species, freed from the natural checks and balances of predation and disease in their home range, to proliferate as invasive species in these new environments. Unchecked, they pose threats to

DOI: 10.1079/9781836991700.0013

the balance of native ecosystems, along with the benefits that these provide to humans. This occurs through competition with native species, habitat modification, predation particularly of early life stages, interbreeding, bringing with them novel diseases and other factors besides.

Topmouth gudgeon

A prominent mention in this 'rogues gallery' of invasive alien small fishes in British waters is the topmouth gudgeon (Fig. 13.1), the Latin name of which is *Pseudorasbora parva* (Temminck & Schlegel, 1846). Topmouth gudgeon are often colloquially referred to as 'TMGs'. They are also known in the aquatic trade as 'stone moroko' or 'clicker barb', the latter name as they can be heard making clicking sounds. TMGs are a kind of gudgeon, as they are members of the gudgeon family (Gobionidae). They are small fish with an elongated, spindle-like body covered in prominent scales weighing up to 16 grams (about half an ounce), with a maximum reported length of 12½ centimetres (5 inches) and a maximum reported age of 5 years. The snout is slender, and the terminal mouth is oriented upwards. A prominent longitudinal pigmented line extends along each flank. The fins of the topmouth gudgeon are not significantly elongated at the base.

The topmouth gudgeon is native to slow-flowing and still fresh waters from Japan to the Amur basin, which is the world's tenth longest river forming the border between the Russian Far East and Manchuria/north-eastern China. Their introduction beyond this range has most likely resulted from releases or escapes from the aquarist trade, but also now occurs through accidental introductions when other fish are transferred between waters.

Their principal diet of invertebrates, fish and fish eggs is a key part of the problem that these small invaders cause when they become established in British and other non-native waters. Though small, topmouth gudgeon have a rapid growth rate, capacity to breed after only their first year, and can spawn multiple times each year on leaves or stones throughout the spring and summer. Topmouth gudgeon also exhibit a degree of parental protection, with male fish guarding the eggs. Topmouth gudgeon can thereby rapidly colonize

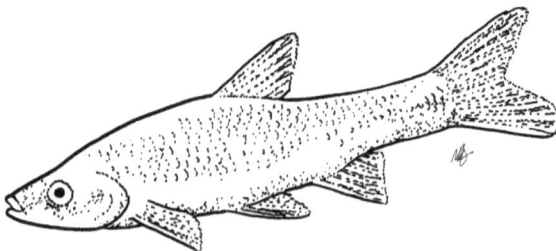

Fig. 13.1. The topmouth gudgeon. (Image © Mark Everard.)

Fig. 13.2. Outline of a topmouth gudgeon. (Image © Mark Everard.)

and proliferate in new waters where, due to their tendency to eat the spawn and fry of other fishes, they both outcompete and prevent the reproduction of native species. Topmouth gudgeon have proliferated to nuisance proportions in some British waters, and active eradication programmes are in place to eliminate them.

The IUCN Red List (The International Union for Conservation of Nature's Red List of Threatened Species) assessed the topmouth gudgeon (Fig. 13.2) as of 'Least Concern' (LC) in terms of its risk of extinction. However, the invasive nature of this fish and its spread into new regions are certainly posing a threat to other fish species.

Sunbleak

The sunbleak (Fig. 13.3), with a Latin name *Leucaspius delineatus* (Heckel, 1843), is a small fish, growing up to 17 grams (just over half an ounce) in weight and a length of up to 9 centimetres (around 3½ inches). Sunbleak are also known as 'belica' or 'motherless minnows', the latter name as the eggs are durable and the fish can appear to arise without parents. They are a shoaling surface-feeding species with flanks coloured bright silver and are readily distinguished by the incomplete lateral line that peters out shortly before the end of the pectoral fin. Sunbleak are native to areas of Asia and Eastern Europe, favouring slow-flowing and still waters.

Non-native to Britain, sunbleak have become locally established in pockets throughout England and are highly invasive. They have also spread widely across Northern Europe initially, it is believed, through releases from the aquatic trade and spreading subsequently through canal systems and

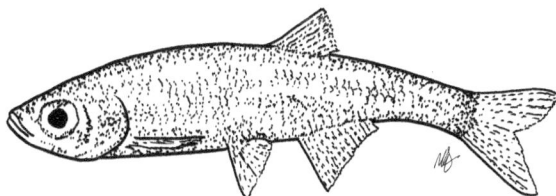

Fig. 13.3. The sunbleak. (Image © Mark Everard.)

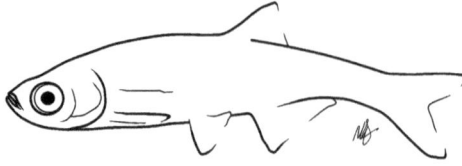

Fig. 13.4. Outline of a sunbleak. (Image © Mark Everard.)

inadvertent movement with other fishery stocks. Sunbleak may also be spread inadvertently as accidental stowaways when other fish are transported, or as eggs adhering to water plants or nets transferred between waters without adequate biosecurity.

Akin to the topmouth gudgeon, sunbleak have a rapid growth rate, are capable of breeding after only their first year and spawn multiple times each year on leaves or stones throughout the spring and summer. Male sunbleak also exhibit parental protection, guarding their own eggs, and this species also tends to eat the spawn and fry of other fishes. Sunbleak therefore tend both to rapidly colonize new waters and to outcompete and prevent the reproduction of native species. As well as thriving in still and slow-moving waters, sunbleak have populated rivers, increasing their threat to native ecosystems and making their long-term eradication all but impossible.

The IUCN Red List assessed the sunbleak as of 'Least Concern' (LC) in terms of its risk of extinction. However, the invasive nature of this fish and its spread into new regions are certainly posing a threat to other fish species. Sunbleak (Fig. 13.4) are also scheduled under Appendix III of the European Union (EU) Bern Convention (on the Conservation of European Wildlife and Natural Habitats 1979). They have also been added to Appendix II of the Bern Convention as "species that are regulated but the exploitation of which is controlled in accordance with the Directive" due to their potential for invasion.

Pumpkinseed

The pumpkinseed (Fig. 13.5), going by the Latin name of *Lepomis gibbosus* (Linnaeus, 1758), is also known as the 'pumpkin-seed sunfish', 'pond perch' or 'sunny'. One of the French names used for this fish is the rather charming 'perche arc-en-ciel', the translation to English of which is 'rainbow perch'. This fish is part of the American sunfish family (Centrarchidae).

Pumpkinseeds are small and brightly coloured predatory fish with a round body profile that is strongly laterally compressed, their shape contributing to the common name 'pumpkinseed', as well as to the name 'panfish' applied to this group of sunfishes in North America where they naturally occur. The

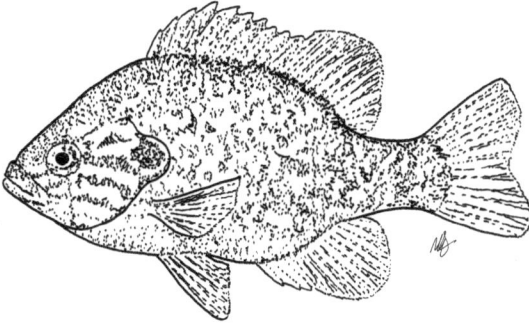

Fig. 13.5. The pumpkinseed. (Image © Mark Everard.)

pumpkinseed has electric-blue highlights across the flanks and strong blue, green and red coloration around the head, with a long dorsal fin consisting of a spined front fin fused into a soft-rayed rear fin. Pumpkinseeds grow up to 40 centimetres (almost 16 inches) long, though more commonly attain a quarter of that length, with a maximum reported age of 12 years.

Pumpkinseeds were introduced into Europe, including the UK, in the 1890s, their bright coloration making them a desirable aquarium fish. Inevitably, there were accidental or deliberate releases into the wild, with scattered populations locally established throughout England and more widely across continental Europe. Some releases are conceivably also from research facilities, as these hardy fish have commonly been used for scientific experiments.

The IUCN Red List assessed the pumpkinseed as of 'Least Concern' (LC) in terms of its risk of extinction. Were this fish more invasive, its spread into new regions could pose a threat to other fish species.

It is on this basis that the pumpkinseed (Fig. 13.6) is scheduled under the UK Wildlife and Countryside Act 1981 (as subsequently amended) within Schedule 9, relating to Section 14 of the Act concerned with 'Introduction of new species etc.'.

Fig. 13.6. Outline of a pumpkinseed. (Image © Mark Everard.)

Bitterling

The European bitterling (Fig. 13.7), a member of the bitterling family (Acheilognathidae), goes by the Latin name of *Rhodeus amarus* (Bloch, 1782). (There was some former confusion of *R. amarus* with *Rhodeus sericeus* native to the Amur River, resolved by a 2006 scientific paper by Jörg Bohlen and colleagues referenced in full in the Bibliography at the end of this book.) There are several species of bitterling across Europe, some of which have also been kept in aquaria, but it is *R. amarus* that concerns us here and the term 'bitterling' will be used in that context.

Bitterling, also referred to as 'bitterling carp', are small fish up to 16 grams (around half an ounce) in weight and reaching a length of 11 centimetres (just under 4½ inches). They have deep, laterally compressed bodies and a short lateral line that peters out five or six scales behind the gill covers. For most of the year, the flanks are generally silver contrasting with the grey-green back, and a distinct metallic stripe extends from the middle of the flank to the base of the tail. The mouth is small and lacks barbels, pointing forwards or slightly to the underside of the blunt snout. Bitterling have an omnivorous diet.

This once-popular aquarium fish has become locally naturalized in the wild in pockets across England with strongholds in still and sluggish waters in south Lancashire, Cheshire, parts of Shropshire and some of the Great Ouse catchment, although nowhere are bitterling abundant in British waters.

The bitterling's favoured habitat is densely weeded regions of still waters and slow-flowing river margins with sandy or muddy bottoms where freshwater mussels occur, mussels playing a key role in the bitterling's unique and fascinating life cycle. In April and May, male bitterling develop a brilliant spawning livery, their flanks taking on a strikingly iridescent hue with the dorsal and anal fins turning bright red. (The Latin name for the genus *Rhodeus* is derived from the Greek 'rhodea', meaning 'rose', referring to the spawning colour of the male's body.) Male bitterling also develop a triangular area of dense white tubercles on either side of the snout during the spring breeding season. At this time, female bitterling grow long, fleshy ovipositors (egg-tubes) extending about 6 centimetres (nearly 2½ inches) from their genital openings. It is at this point

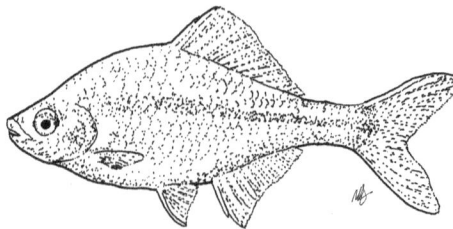

Fig. 13.7. The bitterling. (Image © Mark Everard.)

that the importance of freshwater mussels becomes evident. Male bitterling select a suitable mussel, which they then guard against other male fish. To this, they attract a female bitterling which, if receptive, inserts her ovipositor into the filter-feeding tube of the mussel. She releases one or two eggs down the ovipositor, these eggs sticking to the gills of the mussel where they are protected within the mussel's shell, the male shedding milt into surrounding water that is then drawn into the mantle cavity of the mussel to fertilize the eggs. This behaviour may be repeated with the same or other female bitterling, each female depositing between 40 to 100 eggs, each with a diameter of 3 millimetres (about 0.12 inches). Eggs protected within the mussel's mantle cavity hatch after two or three weeks, the larvae initially remaining within the protective mantle of the mussel while they absorb their yolk sac before then metamorphosing into fry and exiting into the surrounding water environment. In addition to this protective role, mussels also respond to declining water levels, drawing themselves through the sediment into deeper water using their muscular 'foot' and further safeguarding their cargo of bitterling eggs and embryos from drying out.

Bitterling, though an introduced alien species in British waters, have not thus far proven invasive, with no accounts of any significant ecosystem disruption where they are present. None the less, bitterling are scheduled under the UK's Wildlife and Countryside Act 1981 (as subsequently amended) under Schedule 9, Section 14 dealing with 'Introduction of new species etc.'. Bitterling are also included in Appendix III of the Bern Convention, reflecting their potential to become invasive.

The bitterling is assessed under the IUCN Red List as of 'Least Concern' (LC) in terms of its risk of extinction. The EU Habitats Directive (Council Directive 92/43/EEC on the Conservation of Natural Habitats and of Wild Fauna and Flora), addressing the conservation needs of species at risk across their European range, lists bitterling (Fig. 13.8) under Annex II ('Animal and plant species of community interest whose conservation requires the designation of Special Areas of Conservation (SAC)'), though there is an exemption in the UK where the fish is introduced rather than native.

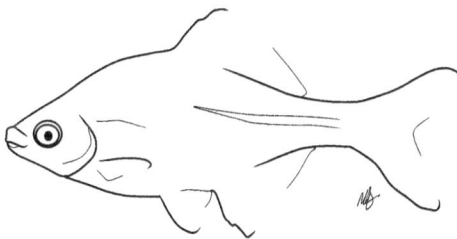

Fig. 13.8. Outline of a bitterling. (Image © Mark Everard.)

Other Unworthy Mentions

In passing, it is worth mentioning some other smaller fishes present, or that have been found, in British waters and that pose a risk of becoming invasive. Much more could be said about these species, but they are addressed here only in summary so as not to divert attention from the other 'small fry' about which this book is principally concerned. In fact, all but one tend potentially to grow larger, and therefore should in the main not strictly qualify as 'small fishes'. They are considered here though as a lesson in the importance of biosecurity.

Goldfish

Prime among this group of alien invasive species is the goldfish, going by the Latin name of *Carassius auratus* (Linnaeus, 1758) and belonging to the true carp family (Cyprinidae). Goldfish are native to central Asia, China and Japan, but have become established widely around the world with reports of adverse ecological impacts in many countries. Not only does the species compete with other fishes where it proliferates, but also it poses a particular risk to the closely related crucian (*Carassius carassius*) with which it hybridizes.

This fish is familiar in its orange and other highly coloured variants in the aquarist trade. However, populations naturalized after release tend to revert to their native general bronze coloration and are generally known as 'brown goldfish'. Goldfish are hardy, widely released and established. Although they may be familiar to many people as small fish, they are, in reality, far too big a species to qualify in these pages with a maximum recorded length of 48 centimetres (19 inches).

Goldfish (Fig. 13.9) appear here more as a warning against the unwise release into the wild of small captive fish, rather than as a qualifying 'small fry' species.

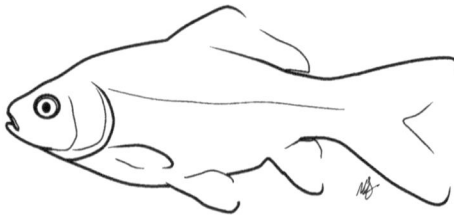

Fig. 13.9. Outline of a goldfish. (Image © Mark Everard.)

Gibel

Gibel, *Carassius gibelio* (Bloch, 1782), also known as 'Prussian carp', are closely related to the crucian (*Carassius carassius*) and also the goldfish, being members of the true carp family Cyprinidae. Gibel occur in still and flowing fresh waters. The natural distribution of gibel is somewhat disputed due to former misidentification with crucians and goldfish, but the fish is usually considered native to central Europe and across to Siberia. However, the fish has been widely introduced across Europe. It is currently absent from the northern Baltic basin, Iceland, Ireland, Scotland and the Mediterranean islands. However, in recent years, populations have become established in southern England, which is a source of great concern.

As gibel can grow up to a recorded length of over 46 centimetres (over 18 inches) they are, in reality, far too big to feature prominently in this book. They are though worthy of mention as they can be problematic, and therefore vigilance and good biosecurity are required to prevent their spread and to avert the threats they pose to native fish populations and ecosystems. This is particularly the case as they are very hardy, tolerating low oxygen conditions and a degree of pollution.

When discussing the spined loach (*Cobitis taenia*), we touched upon the phenomenon of gynogenesis – asexual reproduction where egg development is triggered by sperm but without its DNA fusing with the nucleus of the egg – resulting in the formation effectively of clones. While this may be a rare occurrence in spined loach, it appears to be one of the main methods of reproduction for gibel. Female gibel spawn with several other species, including not just other true carp species but also other cypriniform fishes such roach and common bream. As the eggs develop on contact with sperm without being fertilized, this results in female-only or female-dominated gibel populations, within which males may constitute only a small proportion. This ability to reproduce from unfertilized eggs allows these fish to rapidly populate waters. As Peter Rolfe commented pertinently in his 2023 book *Old Angler Rambling*, "and it only takes one escapee to start a plague". As a recent alien invasive species in Britian, gibel (Fig. 13.10) pose a serious threat.

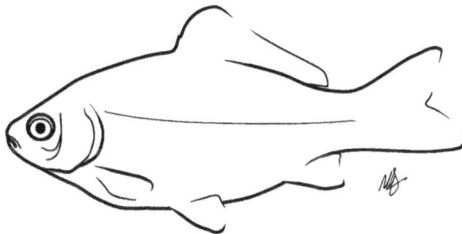

Fig. 13.10. Outline of a gibel. (Image © Mark Everard.)

White sucker

The white sucker, *Catostomus commersonii* (Lacepède, 1803), is a North American fish in the sucker family (Catastomidae). This river fish can also grow quite large, with a maximum reported age of 65 centimetres (about 26 inches) but is generally much smaller. It is noted in the context of this book because a specimen of the white sucker was discovered in 1995 in the River Gade, a tributary of the River Thames in Hertfordshire. Fortunately, no other specimen has since been found, but the Environment Agency is committed to preventing any further introductions as this fish could occupy the same niche as native barbel and chub, competing for food and spawning areas.

The white sucker (Fig. 13.11) is just one 'real-world' example of a potentially invasive fish found in British waters that could be problematic if it becomes established. It serves as a lesson for vigilance and continuing biosecurity to avert future problems.

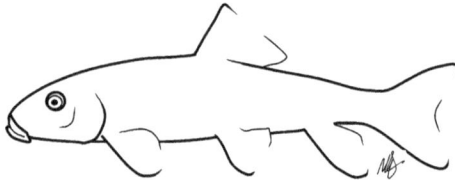

Fig. 13.11. Outline of a white sucker. (Image © Mark Everard.)

American bullheads (catfish)

At least three species of American bullhead catfish are believed to have been imported into the UK for sale through the aquarist trade: American catfish or black bullhead (*Ameiurus melas*); brown bullhead (*Ameiurus nebulosus*); and channel catfish (*Ictalurus punctatus*). All are members of the North American freshwater catfish family (Ictaluridae), possessing eight barbels around the large mouth and strong spines on front edges of the dorsal and pectoral fins, and also lacking scales. They have no relationship with the native bullhead (*Cottus perifretum*).

Of these species, the black bullhead is of the greatest concern in terms of its ability to proliferate rapidly, guarding its eggs, and to compete with native species also consuming their eggs and juveniles.

Though black bullheads can grow relatively large (maximum recorded weight is 3.6 kilograms, or nearly 8 pounds), they more commonly tend to form dense populations of small specimens that shoal in balls for added protection from native predators. They are also known to carry novel parasites and disease. Black bullheads are widely introduced across continental Europe – in France, they are known as 'poisson chat', translating as 'cat fish' – locally forming dense populations and proving problematic.

The only known established black bullheads (Fig. 13.12) in England were in the county of Essex where they formed a dense population in a lake that proved problematic, but this was successfully eradicated in 2014.

Fig. 13.12. Outline of a black bullhead. (Image © Mark Everard.)

Fathead minnow

A single known British population of the American fathead minnow (*Pimephales promelas* Rafinesque, 1820) was eradicated in 2008. The fathead minnow is a small fish, typically reaching a length of 7–10 centimetres (nearly 3 to nearly 4 inches), and is a member of the minnow family (Leuciscidae). It is naturally distributed throughout much of North America from central Canada south along the Rockies to Texas and east to Virginia and the north-eastern USA. The fish though has been introduced to many other regions of the USA, and this is believed to be mainly through its widespread use as a recreational angling baitfish. The torpedo-shaped body is generally coloured dull olive-grey fading to a lighter belly, with a dusky stripe extending along the back and sides and a dusky blotch midway on the dorsal fin (Fig. 13.13).

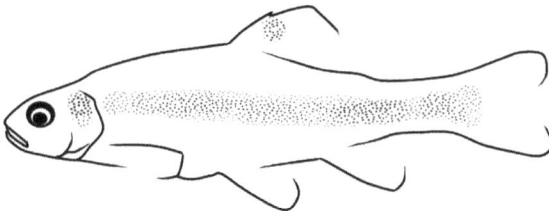

Fig. 13.13. Outline of a fathead minnow. (Image © Mark Everard.)

Controlling the Spread of Small Invaders

The term 'alien invasive' may have science fiction connotations. However, when applied to organisms evolved here on planet Earth, is relates to those that are not native to a given region. Freed from the 'checks and balances' of the ecosystems within which they evolved, these alien organisms can become established and spread, potentially harming the environment, economy and/or human health. A scientific paper published in 1998 specified attributes generally considered to predispose aquatic organisms to become invasive (see Box 13.1).

> **Box 13.1.** Attributes of aquatic organisms predisposed to become invasive
>
> The following attributes were recognized in a scientific paper by Ricciardi and Rasmussen (1998), referenced in the Bibliography at the end of this book, as predisposing aquatic organisms to becoming invasive:
>
> 1. Abundant and widely distributed in their original range.
> 2. Wide environmental tolerance.
> 3. High genetic variability.
> 4. Short generation time.
> 5. Rapid growth.
> 6. Early sexual maturity.
> 7. High reproductive capacity.
> 8. Broad diet (opportunistic feeding).
> 9. Gregariousness.
> 10. Possessing natural mechanisms of rapid dispersal.
> 11. Commensal with human activity (e.g. transport in ship ballast water, or via trade of ornamental species for aquarists).

Some of the smaller freshwater fish species now found in Britain, including topmouth gudgeon, sunbleak, bitterling, pumpkinseed, goldfish and gibel, share some or all of these attributes. So too does the black bullhead, though it is believed that formerly established British populations have now been eradicated. There is, though, a need for constant vigilance and biosecurity if these and other fish species are not to become established and spread, potentially causing harm. It is not just fish species that are of concern: there are many alien invasive invertebrates ('killer shrimp' *Dikerogammarus villosus*, zebra mussels *Dreissena polymorpha* and many more) as well as water plants (such as floating pennywort *Hydrocotyle ranunculoides* and New Zealand swamp stonecrop *Crassula helmsii*) that are now widely found in British waters and causing problems.

Controlling alien invasion by both small and larger species matters a great deal as, when proliferating often at the expense of native species, they can threaten the balance and integrity of our freshwater ecosystems, their wildlife and the many benefits that society derives from them. Alien invasion by fishes and other organisms is a real threat that is often underappreciated and is also frequently irreversible.

It should also be borne in mind that species may be native only to part of country, and that their spread to waters within which they have not evolved can also be disruptive. As we have seen, the 'Doggerland Bridge' established a natural distribution of fish species in the British Isles with geographically impenetrable boundaries between watersheds preventing their spread to the west and north. Equally, lack of marine phases had prevented many fish species from colonizing the island of Ireland.

Many fish species have been translocated, a good number of them with little consequence, but others have caused significant damage. The spread of ruffe into Bassenthwaite in the Lake District, for example, is partly implicated along with agricultural pollution in the extinction of the vendace (*Coregonus albula*) in that English lake. Ecological interactions of ruffe introduced to Scotland's Loch Lomond have also been studied, raising concerns about predation on the eggs of powan (*Coregonus lavaretus*), a fish of national conservation value.

Alien invasive fish and UK law

Threats from alien invasive fish are significant enough to have triggered controlling legislation in the UK. When considering invasive species earlier in this chapter, we have looked at the scheduling of species under the UK Wildlife and Countryside Act 1981 (as subsequently amended) as well as under the EU Bern Convention on the Conservation of European Wildlife and Natural Habitats 1979 and also the EU Habitats Directive. However, two principal pieces of British legislation relate to the potential for fish species to escape and become established.

The Import of Live Fish (England and Wales) Act 1980, applied in England and Wales (with subsequent amendments), controls the release of non-indigenous species of fish. This Act, commonly referred to as ILFA, affords protection to native fish species from invasions. Other than where expressly permitted by licence, it is an offence under ILFA to import, keep or release into the wild any live fish or shellfish (including their eggs or milt) that is not native to England and Wales. Restricted species under ILFA are listed in a schedule to the Act, but there is flexibility under ILFA for the Minister to specify any species "which in the opinion of the Minister might compete with, displace, prey on or harm the habitat of any freshwater fish, shellfish or salmon in England and Wales".

In 1998, the Prohibition of Keeping or Release of Live Fish (Specified Species) Order was introduced in an effort to reduce the number of fishes illegally introduced into fresh waters in England and Wales. This Order lists temperate freshwater fishes believed to pose a threat to the British aquatic environment, specifying that shops have to hold a licence to sell any of the listed species. Fishkeepers are also required to hold a special licence to retain some scheduled fish species in captivity. The 1998 Order also relates to the transfer of fish species used as live bait in recreational angling. Further species have been added to the initial list in the Order (see Box 13.2), but there is also a presumption that the Order should apply to any species that might potentially become established in British waters with unpredictable results. Scotland finalized parallel legislation, publishing The Prohibition of Keeping or Release (Specified Species) Scotland Order 2003.

Box 13.2. Prohibition of Keeping or Release of Live Fish (Specified Species) Order 1998 – Schedule 2: 'Species of fish whose keeping or release in any part of England and Wales is prohibited except under authority of a licence granted by the Minister'

- American brook trout, *Salvelinus fontinalis*
- Asp, *Aspius aspius*
- Big-head carp, *Hypophthalmichthys nobilis* (formerly *Aristichthys nobilis*)
- Bitterling, *Rhodeus sericeus*
- Blageon, *Leuciscus souffia*
- Blue bream, *Abramis ballerus*
- Burbot, *Lota lota*
- Catfish species, of the genera *Ictalurus* and *Silurus*
- Chinese black or snail-eating carp, *Mylopharyngodon piceus*
- Danubian bleak, *Chalcalburnus chalcoides*
- Grass carp, *Ctenopharyngodon idella*
- Landlocked salmon, non-anadromous varieties of the species *Salmo salar*
- Large-mouthed black bass, *Micropterus salmoides*
- Mediterranean barbel, *Barbus meridionalis*
- Nase, *Chondrostoma nasus*
- Pacific salmon and trout (excluding rainbow trout but including steelheads), of the genus *Oncorhynchus*
- Paddlefish, of the genera *Polyodon* and *Psephurus*
- Pike-perch (including zander), of the genus *Stizostedion*
- Pumpkinseed, *Lepomis gibbosus*
- Rock bass, *Ambloplites rupestris*
- Schneider, *Alburnoides bipunctatus*
- Silver carp, *Hypophthalmichthys molitrix*
- Sturgeon or starlet, of the genera *Acipenser*, *Huso*, *Pseudoscaphirhynchus* and *Scaphirhynchus*
- Topmouth gudgeon, *Pseudorasbora parva*
- Toxostome (or French nase), *Chondrostoma toxostoma*
- Vimba, *Vimba vimba*

Why Small Fish Matter

14

If these species are just 'small fry', often referred to dismissively in fishery management circles as 'minor species', can they really be important? The answer to that question is an emphatic YES!

The Roles of 'Small Fry' in Food Webs and Ecosystem Functioning

As described in Chapter 1, the introduction of this book, the great ecosystems of planet Earth are driven not by the massive but by the minute, and particularly the imperceptible. The cumulative biomass of termites on East Africa's great plains, underpinning the dynamics of these tracts of grassland, vastly outweighs the visible biomass of mammals and birds. In the ocean, the cumulative mass of krill substantially outweighs that of mighty giant squid, gargantuan whales and other larger life forms. Just as great trees are fed not merely by their subterranean adventitious rootlets and root hairs but also by the myriad bacteria, fungi and other microscopic organisms associated with their rhizosphere, so it is that the little and the seemingly inconsequential are vastly important for energizing the edifice of ecosystems including their bigger charismatic species. Remove these essential foundations and the whole structure collapses, along with the many ways in which it sustains us bipedal dependents.

Fishes of all sizes play significant roles in many of the chemical and biological cycles within aquatic ecosystems. Through these activities, they contribute to a range of ecosystem services (benefits to people and the maintenance of natural systems) including:

- Provisioning ecosystem services: Directly as sources of food, ornamental resources and genetic resources.
- Regulating ecosystem services: Directly as predators of the vectors of many waterborne diseases as well as potentially vectoring some other

© Mark Everard 2025. *Small Fry: Britain's Tiniest Freshwater Fishes* (M. Everard)
DOI: 10.1079/9781836991700.0014

diseases, spreading seeds through ingestion of plant matter and distribution of propagules in their faeces, and also distributing other animal species such as the glochidia larvae of mussels. They also play various indirect roles in ecosystem processes, purifying water and storing carbon with benefits for climate regulation at global and local scales.

- Cultural ecosystem services: Directly exploited in recreational angling as well as commercial and subsistence fishing, and indirectly as a contributor to the aesthetics, tourism interest and, sometimes, spiritual importance of water bodies. Also serving as a social focus for communities interested in freshwater ecology, wider wildlife and angling.

- Supporting ecosystem services: Playing roles in nutrient, carbon and water recycling, as prey and predator, and as indicators of habitat valuable for a range of wildlife including themselves and other fishes.

For these reasons, water quality standards required to sustain populations of different types of freshwater fish communities are the basis of many strands of environmental legislation that are used to control pollution from industry, domestic sewage and agriculture. The needs of fishes also serve as legal guidelines for the maintenance of minimum water flows and habitat quality.

In fresh waters, as well as marine and terrestrial habitats, it is the 'little guys' that keep global cycles turning, underpinning their resilience and capabilities to deliver a wide range of benefits for people.

'Small Fry' and Recreation

As noted throughout this book, fishing for small fry can be great fun! Significantly too, we have always to recall that the presence and activities of smaller fish species provide a foundation for populations of larger and more generally desirable angling quarry. Take away the 'little guys' and the fishery ecosystem stops functioning.

But recreation is about more than hunting with rod and line. For many a younger angler, going down to the pond or stream with a fishing net and jam jar is, or was, a pursuit that many enjoyed, and from which they gained an enduring connection with and love for the wonders of the natural world.

Recreation too does not have to encompass the capture (and ideally release unharmed) of fish. Simply watching can be enough. Take yourself to a town characterized by a major river – the clean waters of the River Wye in Bakewell (Derbyshire) or the River Test in Stockbridge (Hampshire) come to mind – and visitors are attracted by the sight of trout finning the river below bridges, often eager to intercept pieces of bread, chips and fragments of Bakewell tart thrown into the water. The presence of fish, regardless of whether their observers know what species are involved, as well as other river life that is co-dependent, constitute vital elements of the enriching experience of the riverside.

'Small Fry' and Education

Fish also have a role in education, both formal and informal. In a formal setting, fish serve a range of educational purposes. They can be studied in terms of their roles in food webs; or dissected or otherwise studied for people to learn about morphological adaptations to different life habits. Their behaviour can also be studied for understanding of phenomena such as migration, the forces that drive it and the needs it fulfils. Also, the forms of different fishes, both extant and extinct, are educational about evolution. Study of fish life cycles – from spawning through to egg development, hatching and metamorphosis, shifts in diet with growth through to maturity – also enables people to learn about the changing needs of species over life stages and seasons.

These facets and more can also provide a basis for further education about nature conservation, as well as sustainable development, accounting for such pressures as overfishing, pollution, the impacts of invasive alien species, habitat loss and climate change. These elements of learning can occur in both formal settings – classrooms, lecture halls, continuing professional development – as well as informally via broadcast and other media as well as books, educational apps, magazines and public aquaria.

'Small Fry' and Health

There is increasing interest in healthcare circles in nature prescription. This describes a non-medical intervention encouraging people to connect with the natural world in order to improve their mental and physical health. People are a part of nature, but the pace and demands of modern, urbanizing, increasingly electronic society, reinforced by dense and otherwise poor urban planning, dissociate us from its calming proximity. The term 'nature deficit disorder' (NDD) was coined by Richard Louv in his 2008 book *Last Child in the Woods: Saving Our Children from Nature-Deficit Disorder*, recognizing that modern people, particularly children, now spend less time in contact with nature than in the recent but also evolutionary past, resulting in negative impacts on both physical and mental health. These impacts are compounded by declining attention to the natural world in education and, worse still, a refocusing on nature as a source of risk rather than healthy activity. Consequences, or symptoms, of NDD include not only physical health issues such as obesity but also diminished cognitive abilities, as well as reduced empathy both for other people and the natural environment.

Reconnecting people with nature has been found to improve physical and mental health and to develop empathy. Prescribing time spent in nature for patients is becoming more common, with a proven track record of positive health impacts including stress reduction, socialization, and improved mood, immunity and sleep. Time spent in nature can include fishing, walking, swimming, gardening and other activity as safe and effective interventions, often with other people to promote social connections and shared enriching experiences. Aquarium therapy is another type of intervention wherein people are encouraged to spend time with aquariums, both small at-home aquaria and in larger public facilities, the calming presence of water and fish helping them address any of a range of mental and physical health conditions.

'Small Fry' and Nature Conservation

For all of the benefits noted in this chapter, and for reasons described for each fish species addressed previously in this book, conserving fish species matters a great deal. Some species have been ascribed inherent nature conservation values, and others are noted for their roles in ecosystem processes and resilience. Some are noted as posing conservation threats, including alien invasive species introduced from other countries but also a few that have been problematic when translocated within-country. All play multiple roles, positive or negative, in ecosystem functioning and the broad range of benefits that people derive from healthy ecosystems.

Nature conservation is about far more than altruism or harking back to a former state. It is about the vitality of ecosystems of immense complexity that are readily disrupted with often unforeseen and frequently irreversible consequences. It is also about safeguarding the multiple benefits that people derive from ecosystems, ranging from fresh water and other biophysical resources through to regulation of diseases and the climate, aesthetic and recreational enjoyment, and the integrity of systems vital for supporting livelihoods and lifestyles.

Fish Twitching

There is growing interest in 'fish twitching' (Fig. 14.1). This might sound an odd pursuit, but no one looks askance at birdwatchers and their famed 'bird twitching' habits. Fish twitching does the same for fish.

It is one of the curiosities of human behaviour that attention paid to nature often bounces off the water's surface like so much reflected light. What lies beneath is largely out of sight to many – or even treated as mysterious or possibly even dangerous. Just look at all the negative associations in language associated with 'mires', 'dank places', 'sloughs of despond', 'dark waters' and many more. Yet aquatic ecosystems are varied, diverse and amazingly productive systems important for many reasons. Part of their crowning glories are the many fish species, and life stages of individual species, each adapted to a huge diversity of niches.

Paying more attention to fish-watching, as one of the more charismatic aspects of highly interdependent aquatic life, can open up a wide and perhaps

Fig. 14.1. Polarized glasses and dip net: tools for the keen fish twitcher. (Image © Mark Everard.)

formerly unappreciated vista of nature. A pair of polarizing glasses can help you see better through the surface film by cutting out glare, as can staring down over a bridge parapet to look more vertically through the water's surface. Spend time allowing your eyes to adapt to this new way of looking through, rather than at, the movement of water at the surface, and a panorama is revealed of small bleak, dace and other surface-feeding fish just below the water. Down below, you may suddenly focus on a shoal of minnows, or of gudgeon grazing like so many sheep over a sandy or silty bed. Perhaps a weed bed may roll in the current to reveal the darker and larger form of a barbel, chub or pike. Still waters – ponds or ditches, canals or lakes and reservoirs – can also reveal secrets to the observant watcher. Sticklebacks may be dancing their zigzag dances in the margins, or maybe a ruffe will be hunting down invertebrates, or, with patient observation, a larger fish may roll or otherwise betray its presence.

In pools and streams alike, throwing bread or other offerings on or into the water may attract fish up to feed, revealing them in their burnished glory. Once you gain experience, you may learn to recognize the presence of different species in the dark waters below by the characteristic bubbles that they send to the surface as they grub unseen in the silt on the bed, or you may discern the characteristic rolling behaviour of larger species as they come to the surface to gulp a mouthful of air to charge their swim bladders in order to adjust their buoyancy as they commence feeding.

For some of the more cryptic or elusive little fish species, a more active approach is required. This need not be onerous. A simple pond net worked in the margins of the water can reveal many hidden gems: fishy, amphibian and insect alike. For others, such as bullheads and stone loach, turning over stones and woody debris may be required but always, of course, replacing these 'caves' back exactly as you found them. Also, dabbling in the soft silt in river margins may reveal more cryptic species such as the ammocoete larvae of brook lampreys.

A whole fascinating world is there to be discovered by the fish twitcher. At the very least, it is a harmless occupation. But, as a means to reconnect us with the everyday wonders of the natural world, it is so much more.

Collins Guide to
The Freshwater Fishes
Britain and Europe
The Bristol Avon
Fish, Freshwater Life & Fishing
FISH
IZAAK WALTON
The Compleat Angler
An EVERYMAN Paperback
THE COMPLEX
LIVES OF BRITISH
FRESHWATER
FISHES
SMALL FRY
HOW TO CATCH THEM
KENNETH MANSFIELD
MARK EVERARD
THE LITTLE BOOK OF
LITTLE
FISHES
The spiky little
freshwater ruffian

Bibliography

The following works are referenced in this book, with my thanks to the authors concerned where quoted.

Adams, C.E. and Maitland, P.S. (1998) The ruffe population of Loch Lomond, Scotland: its introduction, population expansion, and interaction with native species. *Journal of Great Lakes Research* 24(2), 249–262. https://doi.org/10.1016/S0380-1330(98)70817-2

'BB' (Watkins-Pitchford, D.J.) (1946) *The Fisherman's Bedside Book*. Eyre & Spottiswoode, London. (Reprinted by Merlin Unwin Books, Ludlow, 2004.)

'BB' (Watkins-Pitchford, D.J.) (1987) *Fisherman's Folly*, revised edition. Boydell Press. Woodbridge, Suffolk.

Bell, A.P. (1926) *Fresh-water Fishing for the Beginner*. Warne's Recreation Books, Frederick Warne & Co. Ltd., London.

Berners, J. (1496) *Treatyse of Fysshynge with an Angle*. Wynkyn de Worde of Westminster.

Bohlen, J., Šlechtová, V., Bogutskaya, N. and Freyhof, J. (2006) Across Siberia and over Europe: phylogenetic relationships of the freshwater fish genus *Rhodeus* in Europe and the phylogenetic position of *R. sericeus* from the River Amur. *Molecular Phylogenetics and Evolution* 40(3), 856–865. https://doi.org/10.1016/j.ympev.2006.04.020

Buckland, F.T. (1873) *Familiar History of British Fishes*. Society for Promoting Christian Knowledge, London.

Campbell, A. and Dawes, J. (eds) (2005) Fish, What is a? In: *Encyclopedia of Underwater Life: Aquatic Invertebrates and Fishes*. Oxford University Press, Oxford.

Canal and River Trust (n.d.) *It's a ruffe old world out there*. Canal and River Trust, Ellesmere Port, UK. Available at: https://canalrivertrust.org.uk/enjoy-the-waterways/fishing/related-articles/the-fisheries-and-angling-team/its-a-ruffe-old-world-out-there (accessed 28 April 2025).

Carim, K.J., Larson, D.C., Helstab, J.M., Young, M.K. and Docker, M.F. (2023) A revised taxonomy and estimate of species diversity for western North American *Lampetra*. *Environmental Biology of Fishes* 106, 817–836. https://doi.org/10.1007/s10641-023-01397-y

Carpenter-Bundhoo, L. and Moffatt, D.B. (2024) Expanding the known range and practical conservation issues of the Endangered Australian brook lamprey *Mordacia praecox*. *Endangered Species Research* 53, 547–553. https://doi.org/10.3354/esr01319

Coll-Costa, C., Dahms, C., Kemppainen, P., Alexandre, C.M., Ribeiro, F., *et al.* (2024) Parallel evolution despite low genetic diversity in three-spined sticklebacks. *Proceedings of the Royal Society B: Biological Sciences* 291, 20232617. https://doi.org/10.1098/rspb.2023.2617

Copp, G.H., Carter, M.G., England, J. and Britton, J.R. (2006) Reoccurrence of the white sucker *Catostomus commersonii* in the River Gade (Hertfordshire). *The London Naturalist* 85, 115–119.

Coxon, H. (1896) *A Modern Treatise on Practical Coarse Fish Angling: How to Catch Fish*. Charles H. Richards, Nottingham, UK. (Republished by The Medlar Press, Ellesmere, UK, 2004.)

Davies, D., Shelley, J. and Harding, P. (eds) (2004) *Freshwater Fishes in Britain: The Species and Their Distribution*. Harley Books, London.

De Cahsan, B., Nagel, R., Schedina, I.-M., King, J.K., Bianco, P.G., *et al.* (2020) Phylogeography of the European brook lamprey (*Lampetra planeri*) and the European river lamprey (*Lampetra fluviatilis*) species pair based on mitochondrial data. *Journal of Fish Biology* 96(4), 905–912. https://doi.org/10.1111/jfb.14279

Dennys, J. (1613) *Secrets of Angling*. London.

Docker, M.F., Youson, J.H., Beamish, R.J. and Devlin, R.H. (1999) Phylogeny of the lamprey genus *Lampetra* inferred from mitochondrial cytochrome b and ND3 gene sequences. *Canadian Journal of Fisheries and Aquatic Sciences* 56(12), 2340–2349. https://doi.org/10.1139/f99-171

Eschmeyer's Catalog of Fishes (online version, updated 1 March 2021) Institute for Biodiversity Science and Sustainability, California Academy of Sciences, San Francisco, California. Available at: https://www.calacademy.org/scientists/projects/eschmeyers-catalog-of-fishes (accessed 28 April 2025).

Espanhol, R., Almeida, P.R., Alves, M.J. and Alves, M.J. (2007) Evolutionary history of lamprey paired species *Lampetra fluviatilis* (L.) and *Lampetra planeri* (Bloch) as inferred from mitochondrial DNA variation. *Molecular Ecology* 16(9), 1909–1924. https://doi.org/10.1111/j.1365-294X.2007.03279.x

Everard, M. (2008) *The Little Book of Little Fishes*. The Medlar Press, Ellesmere, UK.

Everard, M. (2020) *The Complex Lives of British Freshwater Fishes*. CRC/Taylor and Francis, Boca Raton, Florida and Abingdon, UK.

Everard, M. (2023a) *Gudgeon: The Angler's Favourite Tiddler*. CRC/Taylor and Francis, Boca Raton, Florida and Abingdon, UK.

Everard, M. (2023b) *Ruffe: The Spiky Freshwater Ruffian*. CRC/Taylor and Francis, Boca Raton, Florida and Abingdon, UK.

Everard, M. and Pickett, J. (2025) The European bullhead in Britain: a case of mistaken identity. *British Wildlife* 36(4), 266–271.

Franck, R. (1694) *Northern Memoirs*. London.

Freyhof, J. (1999) Die Schmerlen der Gattung Cobitis Eine verwirrende Artengruppe Steinbeisser [Loaches of the genus *Cobitis*: a puzzling species group spined loaches]. *DATZ* 52(11), 14–18.

Freyhof, J., Kottelat, M. and Nolte, A. (2005) Taxonomic diversity of European *Cottus* with description of eight new species (Teleostei: Cottidae). *Ichthyological Exploration of Freshwaters* 16(2), 107–172.

Garnett, D. and Mytton, A. (2019) *Hooked on Lure Fishing*. Merlin Unwin Books Ltd, Ludlow, UK.

Giles, N. (1994) *Freshwater Fish of the British Isles: A Guide for Anglers and Naturalists*. Swan Hill Press, Shrewsbury, UK.

Giraldus Cambrensis (1187) *The History and Topography of Ireland*. (Republished by John J. O'Meara, Penguin, Harmondsworth, UK, 1982.)

Gutsch, M. and Hoffman, J. (2016) A review of ruffe (*Gymnocephalus cernua*) life history in its native versus non-native range. *Reviews in Fish Biology and Fisheries* 26, 213–233. https://doi.org/10.1007/s11160-016-9422-5

Hall, Mr and Mrs S.C. (1859) *The Book of the Thames From Its Rise to Its Fall*. Charlotte James, London.

HM Government (1980) *Import of Live Fish (England and Wales) Act 1980*. HM Government, London. Available at: https://www.legislation.gov.uk/ukpga/1980/27 (accessed 28 April 2025).

HM Government (1998) *Prohibition of Keeping or Release of Live Fish (Specified Species) Order 1998*. STATUTORY INSTRUMENTS 1998 No. 2409. HM Government, London. Available at: https://www.legislation.gov.uk/uksi/1998/2409/made (accessed 28 April 2025).

Houghton, W. (1879) *British Fresh-Water Fishes*. William Mackenzie, London. (Note: This book has been reprinted over the decades by numerous publishers, e.g. by The Peerage Press, London, 1981.)

Hwang, D.-S., Byeon, H.K. and Lee, J.-S. (2013) Complete mitochondrial genome of the river lamprey, *Lampetra japonica* (Petromyzontiformes, Petromyzonidae). *Mitochondrial DNA* 24(4), 406–408. https://doi.org/10.3109/19401736.2013.763246

Keene, J.H. (1881) *The Practical Fisherman: Dealing with the Natural History, the Legendary Lore, the Capture of British Freshwater Fish, and Tackle and Tackle Making*. Bazaar Office, London.

Louv, R. (2008) *Last Child in the Woods: Saving Our Children from Nature-Deficit Disorder*. Algonquin Books, Chapel Hill, North Carolina.

Maitland, P.S. and Campbell, R.N. (1992) *Freshwater Fishes of the British Isles*. HarperCollins Publishers, London.

Mansfield, K. (1958) *Small Fry and Bait Fishes: How to Catch Them*. Herbert Jenkins Limited, London.

Marshall-Hardy, E. (1943) *Coarse Fish*. Herbert Jenkins Limited, London.

Mateus, C.S., Stange, M., Berner, D., Roesti, M., Quintella, B.R., *et al.* (2013) Strong genome-wide divergence between sympatric European river and brook lampreys. *Current Biology* 23(15), R649–R650. https://doi.org/10.1016/j.cub.2013.06.026

Muus, B.J. and Dahlstrom, P. (1967) *Collins Guide to the Freshwater Fishes of Britain and Europe*. Collins, London.

NNSS (n.d.) *Risk Assessment Summary Sheet: Black bullhead (*Ameiurus melas*)*. GB Non-Native Species Secretariat (NNSS). Available at: https://www.nonnatives-pecies.org/assets/Uploads/Ameiurus_melas_black_bullhead_RRA.pdf (accessed 28 April 2025).

Pereira, A.M., Almada, V.C. and Doadrio, I. (2011) Genetic relationships of brook lamprey of the genus *Lampetra* in a Pyrenean stream in Spain. *Ichthyological Research* 58, 278–282. https://doi.org/10.1007/s10228-011-0218-2

Pinder, A.C. (2001) *Keys to Larval and Juvenile Stages of Coarse Fishes from Fresh Waters in the British Isles*. Freshwater Biological Association Scientific Publications Volume 60. Freshwater Biological Association, Windermere, UK.

Potter, B. (1906) *The Tale of Mr. Jeremy Fisher*. Frederick Warne & Co. Ltd, London.

Ransome, A. (1929) *Rod and Line*. Jonathan Cape, London.

Ricciardi, A. and Rasmussen, J.B. (1998) Predicting the identity and impact of future biological invaders: a priority for aquatic resource management. *Canadian Journal of Fisheries and Aquatic Sciences* 55, 1759–1765. https://doi.org/10.1139/f98-066

Robertson, H.R. (1875) *Life on the Upper Thames*. Virtue, Spalding, & Co., London.

Rolfe, P. (2023) *Old Angler Rambling*. Kindle Direct Publishing.

Scharpf, C. (2024) Order Perciformes: Suborder Cottoidea: Infraorder Cottales: Family Cottidae sculpins. *The ETYFish Project*, v. 10.0 – 1 Nov. 2024. Available at: https://etyfish.org/perciformes20/ (accessed 28 April 2025).

Schönhuth, S., Vukić, J., Šanda, R., Yang, L. and Mayden, R.L. (2018) Phylogenetic relationships and classification of the Holarctic family Leuciscidae (Cypriniformes: Cyprinoidei). *Molecular Phylogenetics and Evolution* 127, 781–799. https://doi.org/10.1016/j.ympev.2018.06.026

Souissi, A., Besnard, A.L. and Evanno, G. (2022) A SNP marker to discriminate the European brook lamprey (*Lampetra planeri*), river lamprey (*L. fluviatilis*) and their hybrids. *Molecular Biology Reports* 49, 10115–10119. https://doi.org/10.1007/s11033-022-07800-8

Sterba, G. (1962) *Freshwater Fishes of the World*. Vista Books, London.

Svanberg, I. and Locker, A. (2020) Caviar, soup and other dishes made of Eurasian ruffe, *Gymnocephalus cernua* (Linnaeus, 1758): forgotten foodstuff in central, north and west Europe and its possible revival. *Journal of Ethnic Foods* 7, 3. https://doi.org/10.1186/s42779-019-0042-2

Torresen, O.K., Garmann-Aarhus, B., Hoff, S.N.K., Jentoft, S., Svensson, M., *et al.* (2025) Comparison of whole-genome assemblies of European river lamprey (*Lampetra fluviatilis*) and brook lamprey (*Lampetra planeri*). *bioRxiv*, preprint posted 24 April. https://doi.org/10.1101/2024.12.06.627158

Walton, I. and Cotton, C. (1653) *The Compleat Angler*. Maurice Clark, London. (Available these days in many editions and from various publishers.)

Wells, A.L. (1941) *The Observer's Book of Freshwater Fishes of the British Isles*. Frederick Warne & Co. Ltd, London.

Wheeler, A. (1969) *The Fishes of the British Isles and North-West Europe*. Macmillan, London.

Wheeler, A. (1978) Hybrids of bleak, *Alburnus alburnus*, and chub, *Leuciscus cephalus* in English rivers. *Journal of Fish Biology* 13(4), 467–473. https://doi.org/10.1111/j.1095-8649.1978.tb03456.x

Appendix: Reported Distribution of Small Freshwater Fishes Across Britain

This Appendix comprises distribution maps across Britain, where available, for the small freshwater fish species listed in this book. These distribution maps were kindly supplied by the Biological Records Centre (BRC) on behalf of the Freshwater Fish Recording Scheme. BRC receives support from the Joint Nature Conservation Committee and the UK Centre for Ecology & Hydrology (via the Natural Environment Research Council award number NE/Y006208/1 as part of the NC-UK programme delivering National Capability). I would like to thank Colin Harrower (BRC) for supplying maps. The BRC is indebted to volunteer recorders and organizations who provide data to the scheme, including a substantial contribution of records from the Environment Agency.

These maps use the data from the Database & Atlas of Freshwater Fishes (DAFF) published in the book *Freshwater Fishes in Great Britain* (Davies *et al.*, 2004) with additional records from iRecord data that have been accepted or assumed to be correct (not rejected and or has not been queried though some may not have yet undergone expert verification). Regrettably, the DAFF database and iRecord lack data for these species from Ireland, which is therefore omitted from maps that cover only England, Scotland, Wales and the Channel Islands.

Some iRecord data for pumpkinseed (*Lepomis gibbosus*) have been removed due to potential confusion with a marine species. For white sucker (*Catostomus commersonii*), the data source is a paper by Copp *et al.* (2006) recording that this North American species had only once been reported in open waters of Europe (a single specimen found in the River Gade at Hemel Hempstead, Hertfordshire, in 1992) but was found again in 2004 when an additional seven specimens were discovered during routine and follow-up surveys.

The following time periods were selected to show changes over time with associated colour codings:

- < 1900
- 1900–1949
- 1950–1969
- 1970–1989
- 1990–2009
- 2010 +

Distribution maps (Figs A1.1–A1.20) are sequenced according to the order in which these fishes appear in the body of this book, commencing with cyprinid fishes.

Minnow Distribution

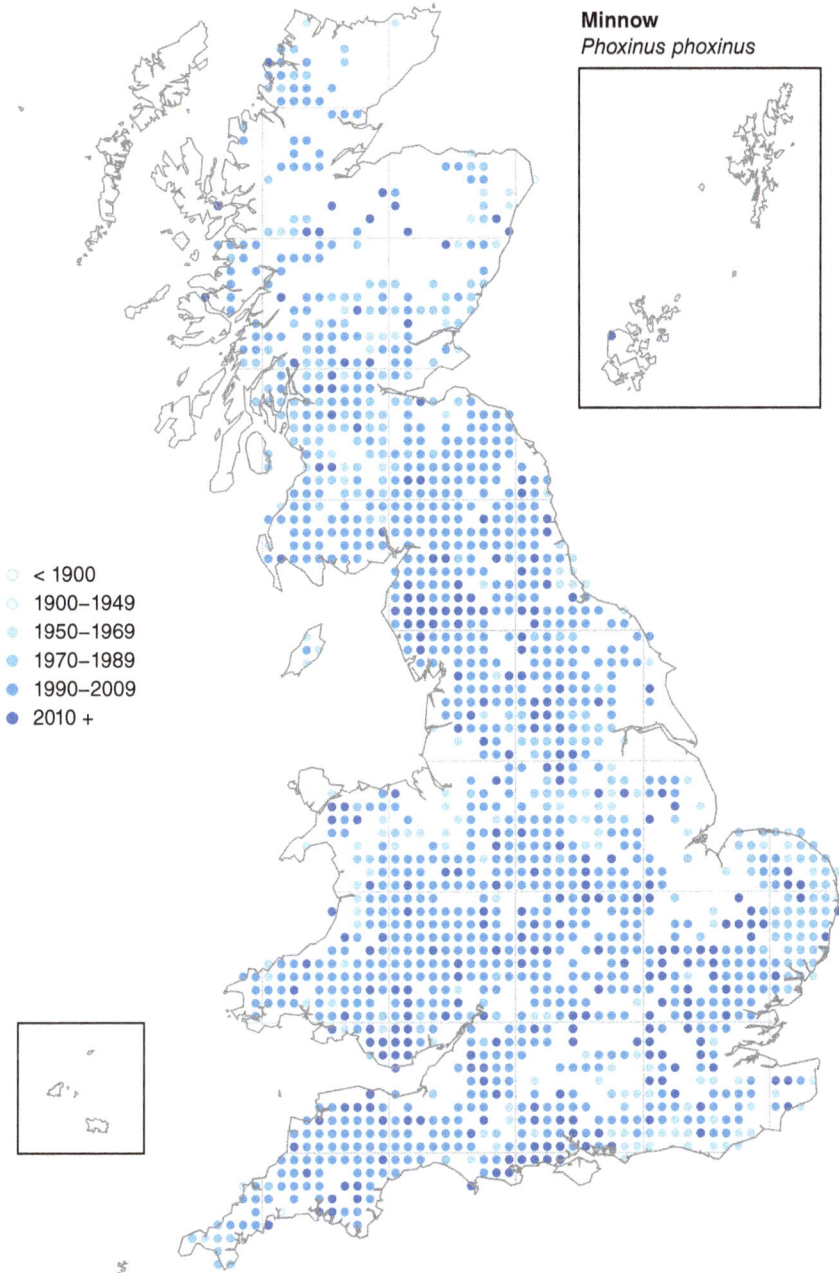

Minnow
Phoxinus phoxinus

○ < 1900
◐ 1900–1949
◑ 1950–1969
● 1970–1989
● 1990–2009
● 2010 +

Fig. A1.1. Reported distribution of minnow across Britain (excluding Ireland).

Gudgeon Distribution

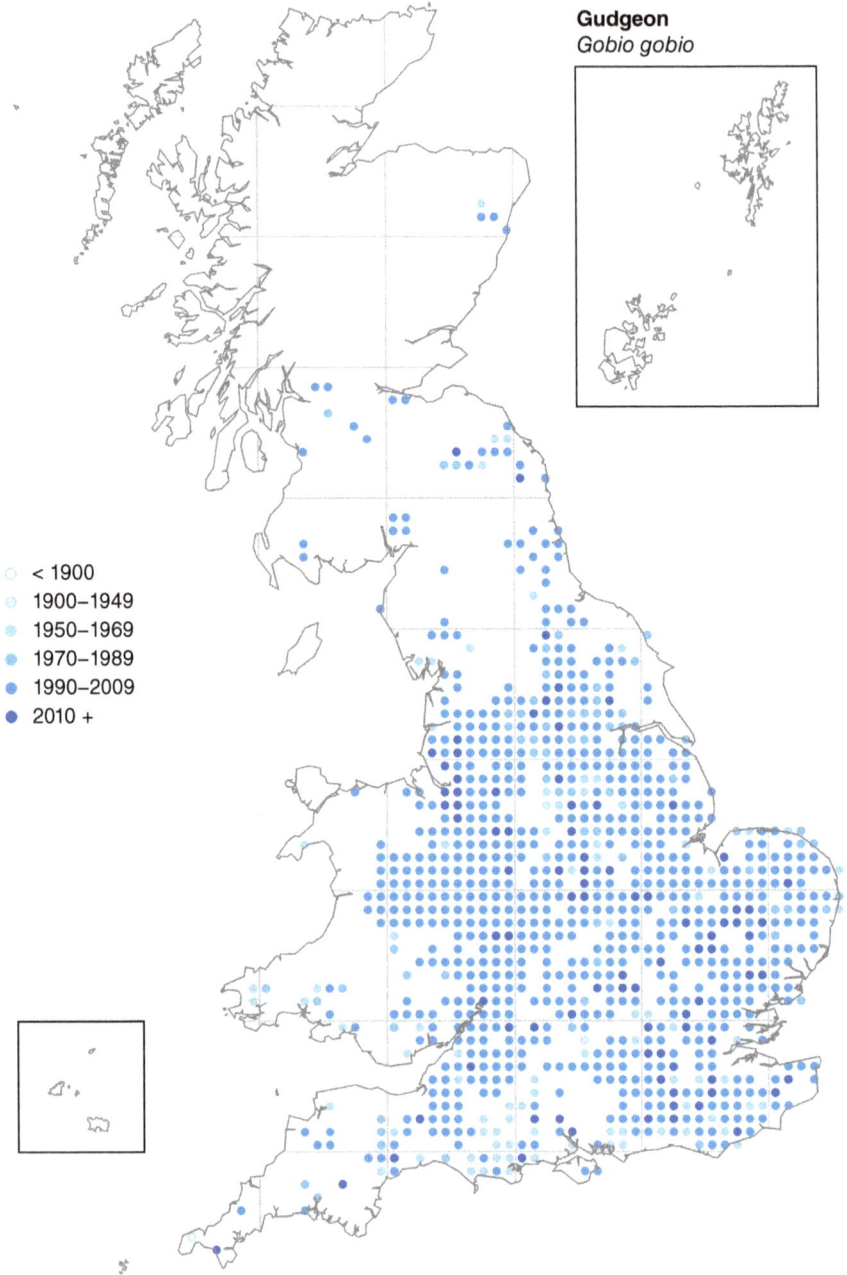

Gudgeon
Gobio gobio

○ < 1900
◐ 1900–1949
◑ 1950–1969
● 1970–1989
● 1990–2009
● 2010 +

Fig. A1.2. Reported distribution of gudgeon across Britain (excluding Ireland).

Bleak Distribution

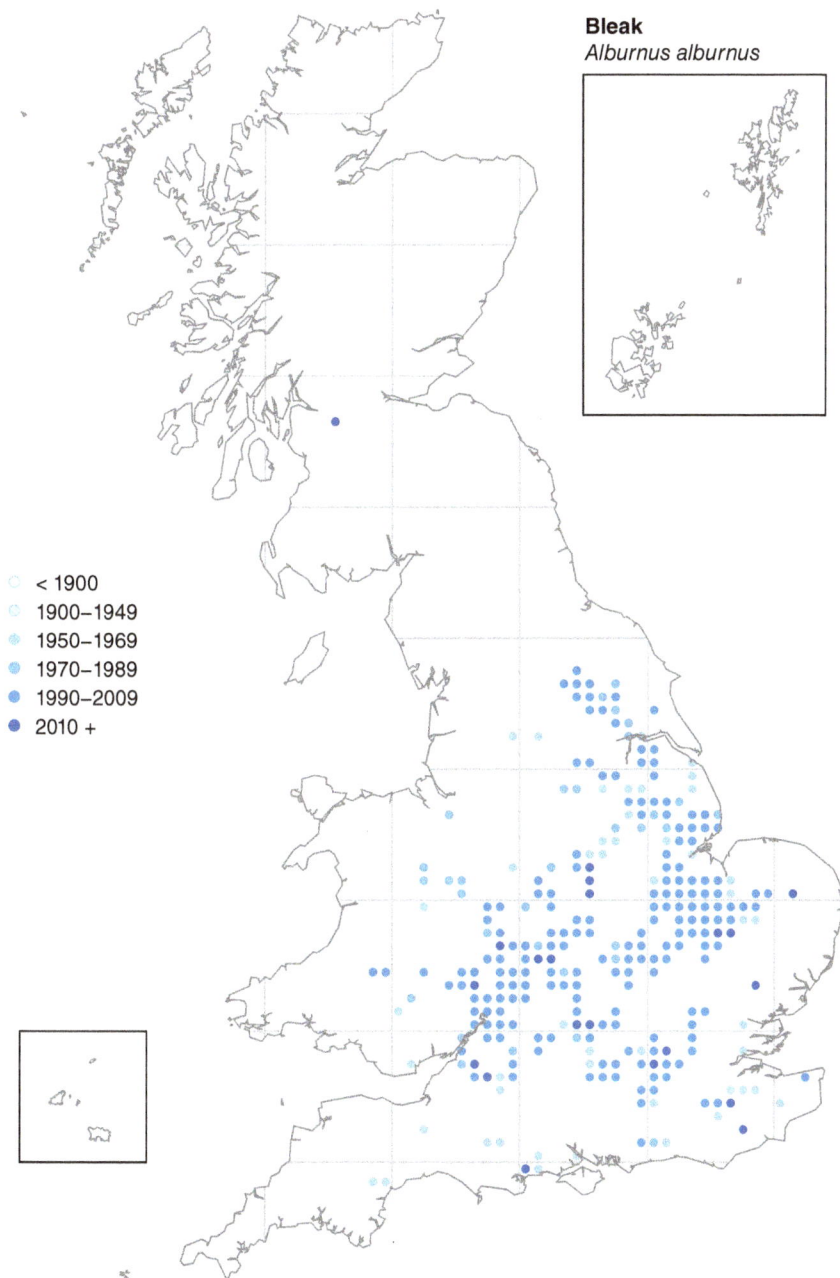

Bleak
Alburnus alburnus

○ < 1900
○ 1900–1949
◑ 1950–1969
◉ 1970–1989
● 1990–2009
● 2010 +

Fig. A1.3. Reported distribution of bleak across Britain (excluding Ireland).

Stone Loach Distribution

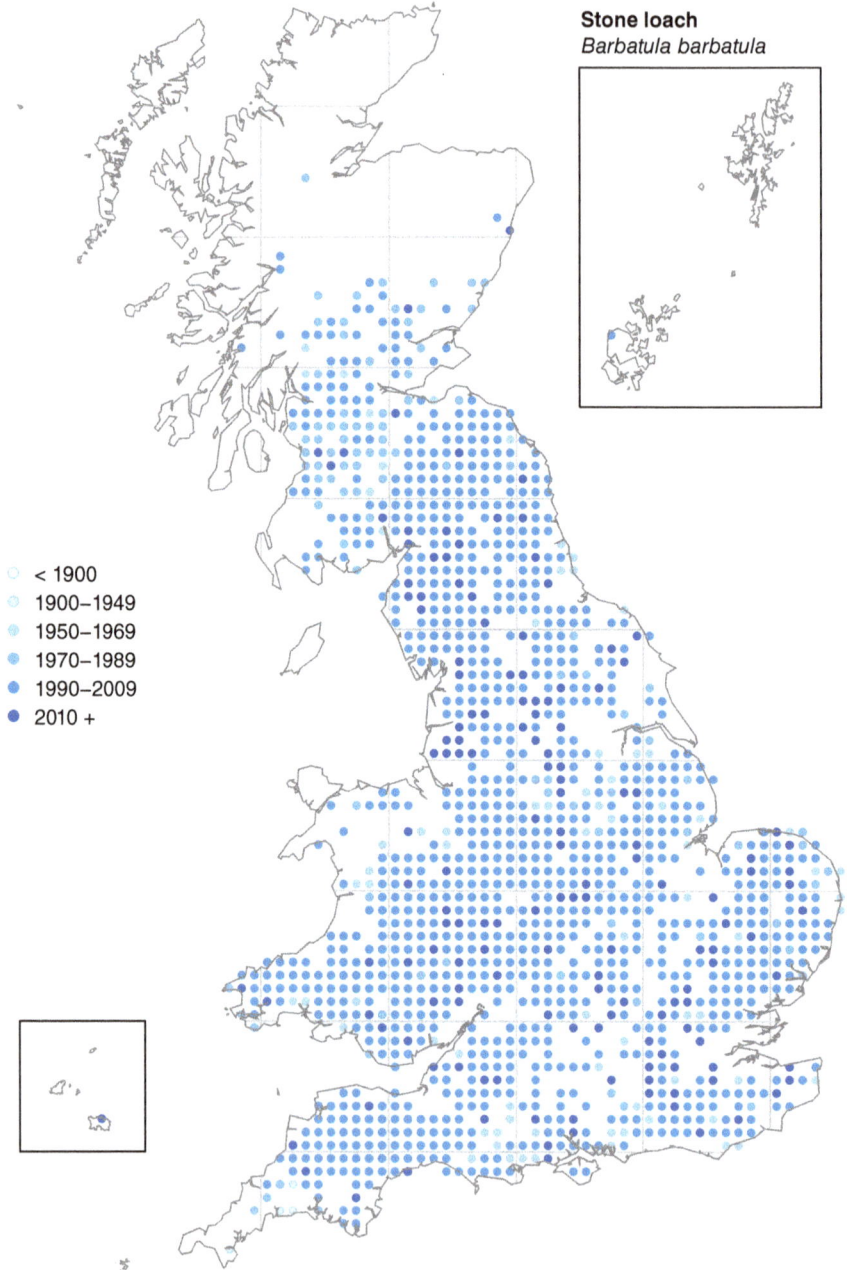

Stone loach
Barbatula barbatula

○ < 1900
○ 1900–1949
◐ 1950–1969
● 1970–1989
● 1990–2009
● 2010 +

Fig. A1.4. Reported distribution of stone loach across Britain (excluding Ireland).

Spined Loach Distribution

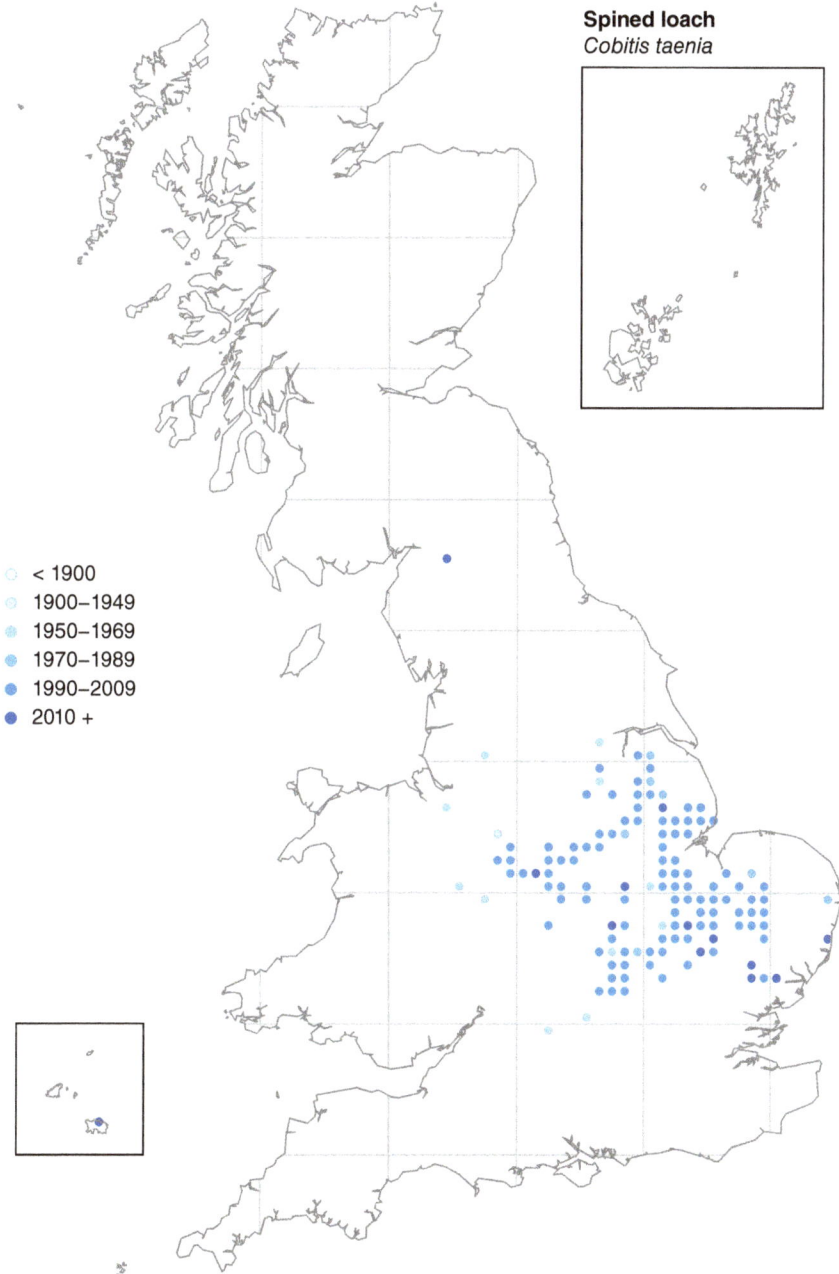

Spined loach
Cobitis taenia

○	< 1900
◔	1900–1949
◑	1950–1969
◕	1970–1989
●	1990–2009
●	2010 +

Fig. A1.5. Reported distribution of spined loach across Britain (excluding Ireland).

Bullhead Distribution

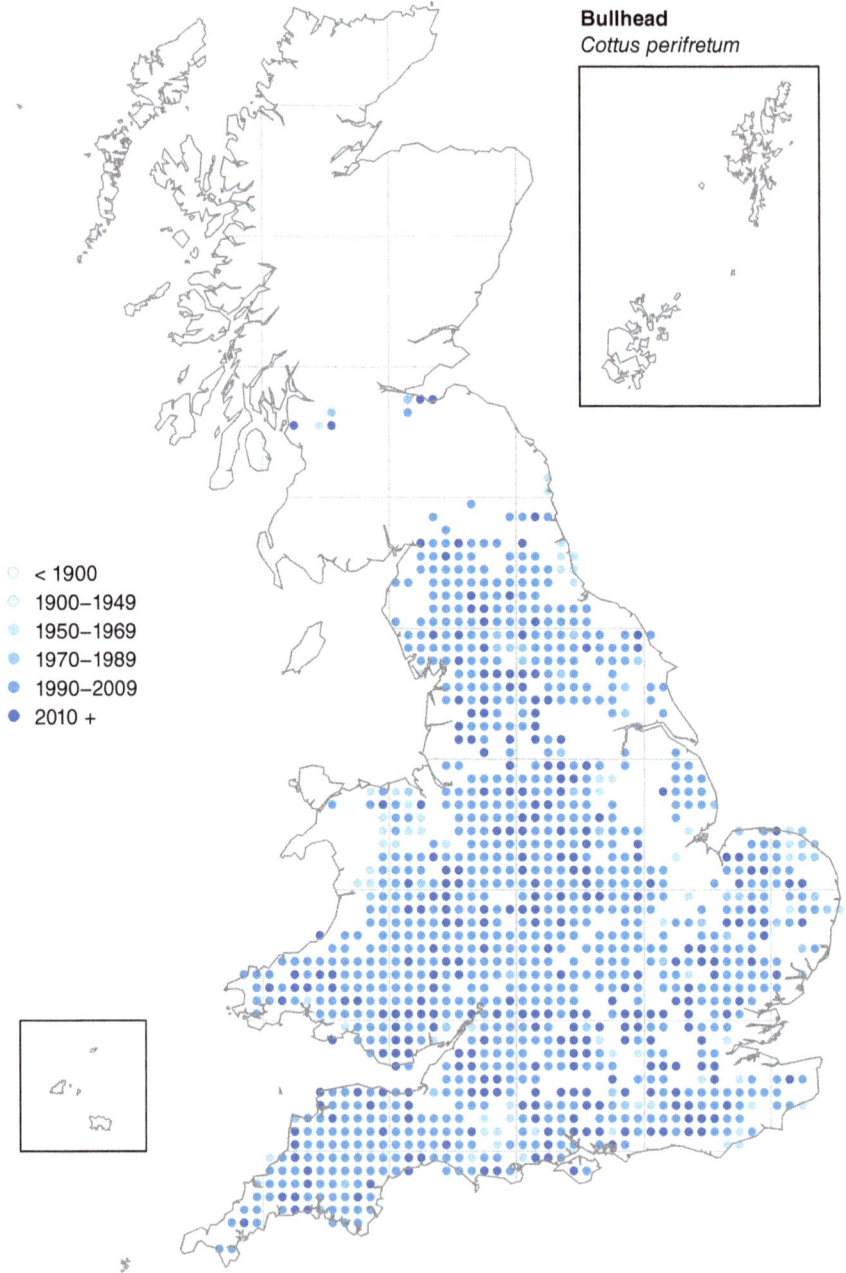

Bullhead
Cottus perifretum

○ < 1900
◐ 1900–1949
◑ 1950–1969
◒ 1970–1989
● 1990–2009
● 2010 +

Fig. A1.6. Reported distribution of bullhead across Britain (excluding Ireland).

Three-Spined Stickleback Distribution

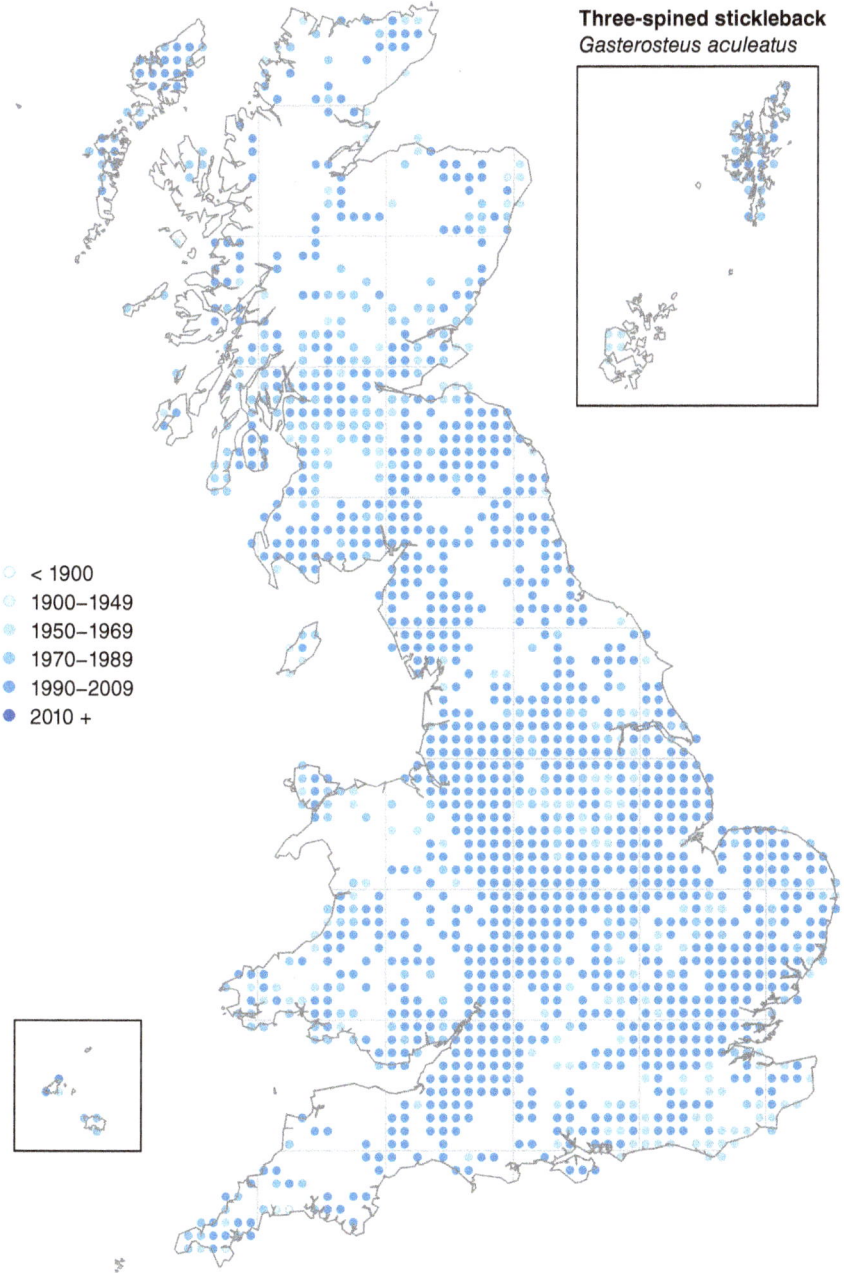

Three-spined stickleback
Gasterosteus aculeatus

○ < 1900
○ 1900–1949
● 1950–1969
● 1970–1989
● 1990–2009
● 2010 +

Fig. A1.7. Reported distribution of three-spined stickleback across Britain (excluding Ireland).

Ten-Spined Stickleback Distribution

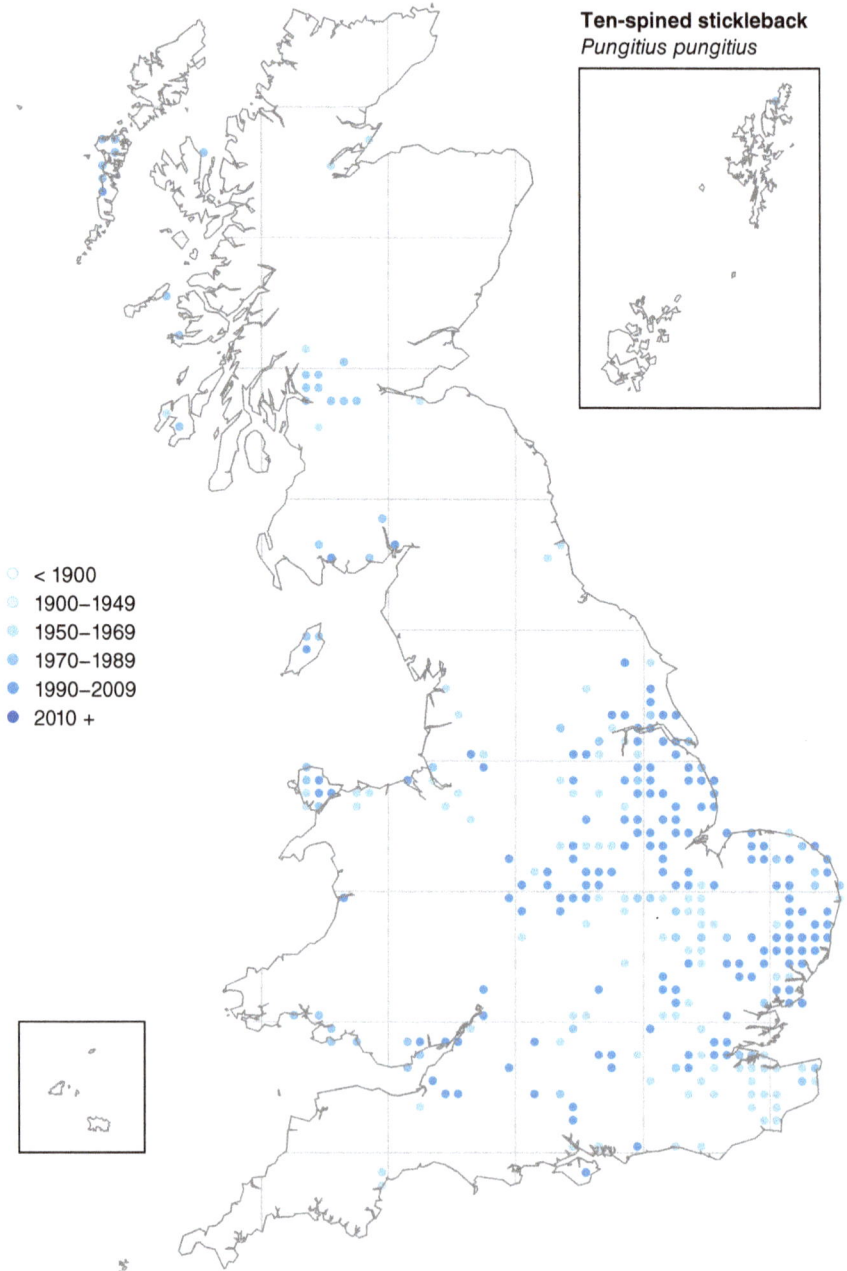

Ten-spined stickleback
Pungitius pungitius

○ < 1900
◔ 1900–1949
◑ 1950–1969
◕ 1970–1989
● 1990–2009
● 2010 +

Fig. A1.8. Reported distribution of ten-spined stickleback across Britain (excluding Ireland).

Ruffe Distribution

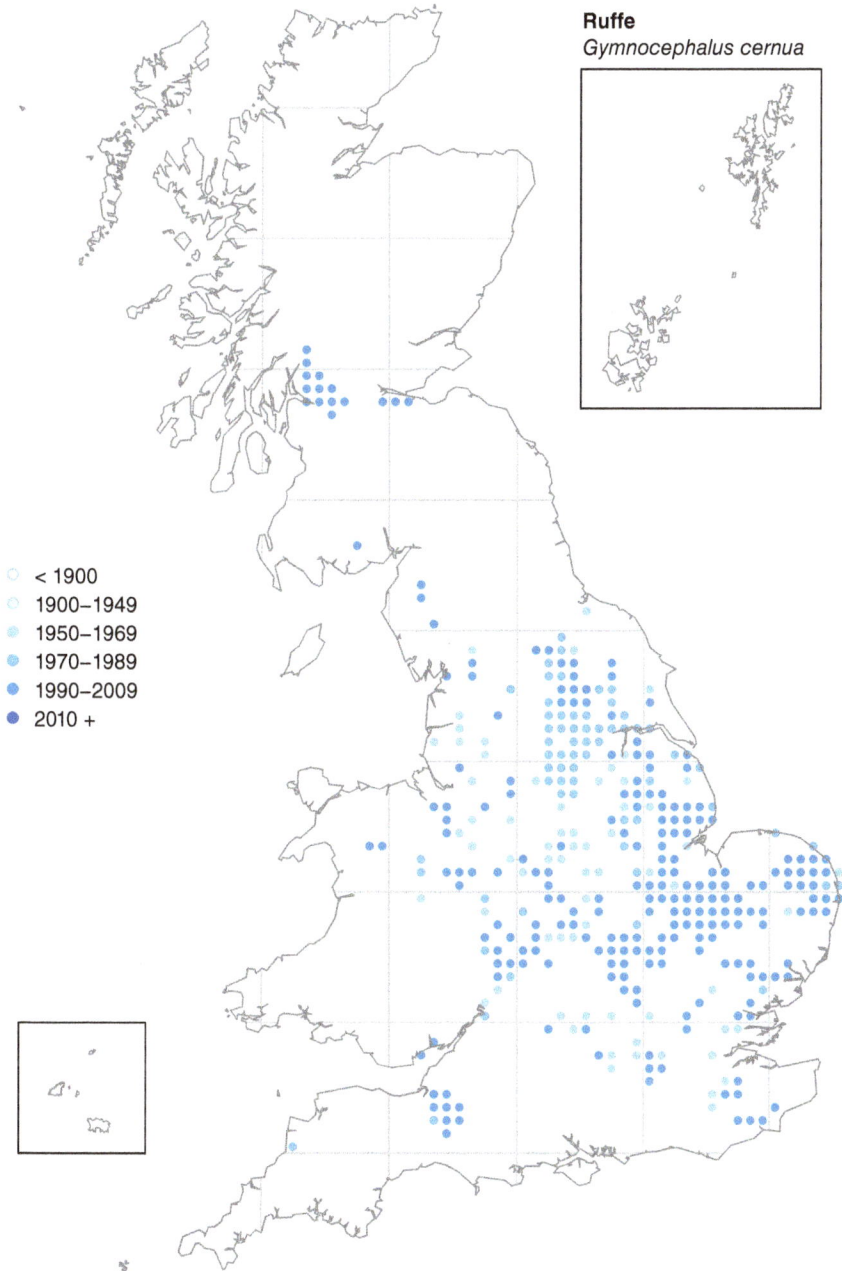

Ruffe
Gymnocephalus cernua

○ < 1900
○ 1900–1949
◐ 1950–1969
◑ 1970–1989
● 1990–2009
● 2010 +

Fig. A1.9. Reported distribution of ruffe across Britain (excluding Ireland).

Brook Lamprey Distribution

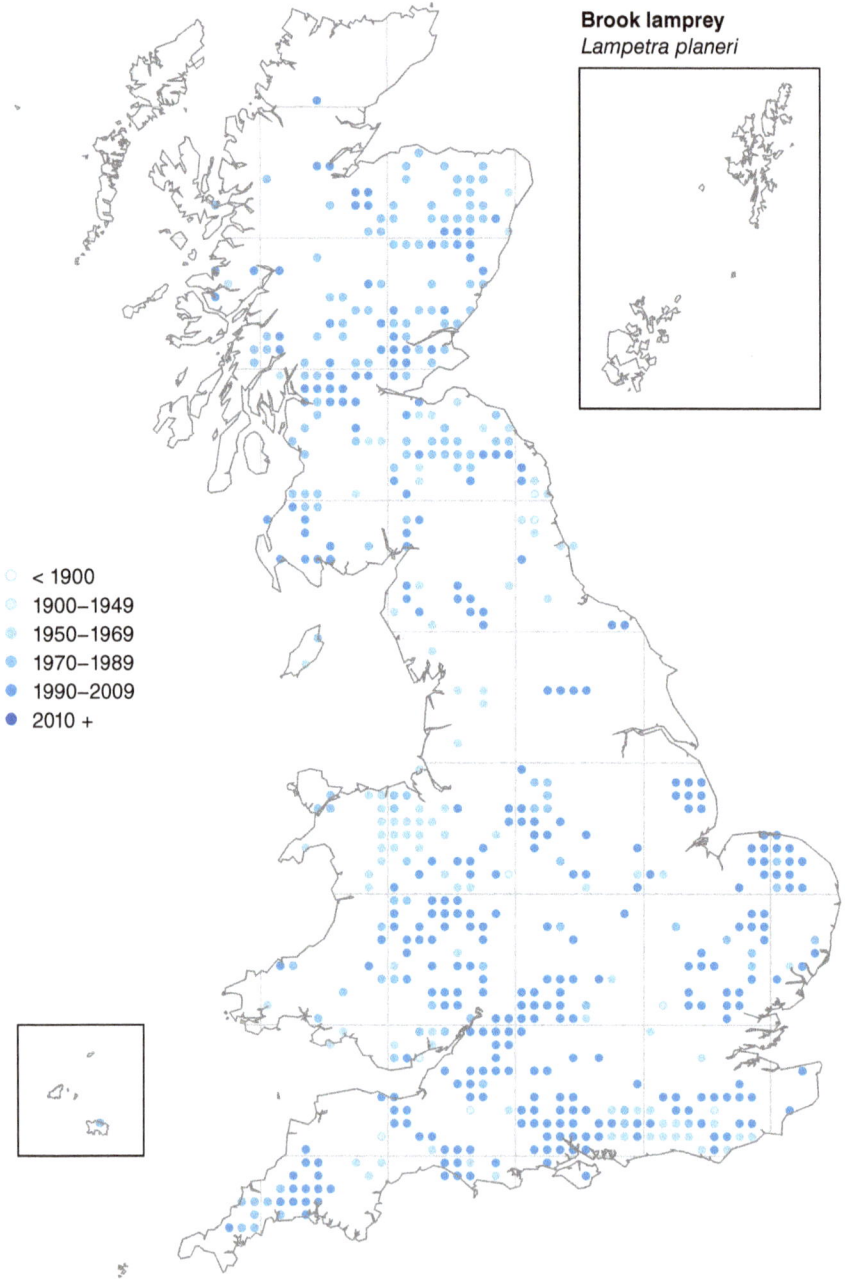

Brook lamprey
Lampetra planeri

< 1900
1900–1949
1950–1969
1970–1989
1990–2009
2010 +

Fig. A1.10. Reported distribution of brook lamprey across Britain (excluding Ireland).

Smelt Distribution

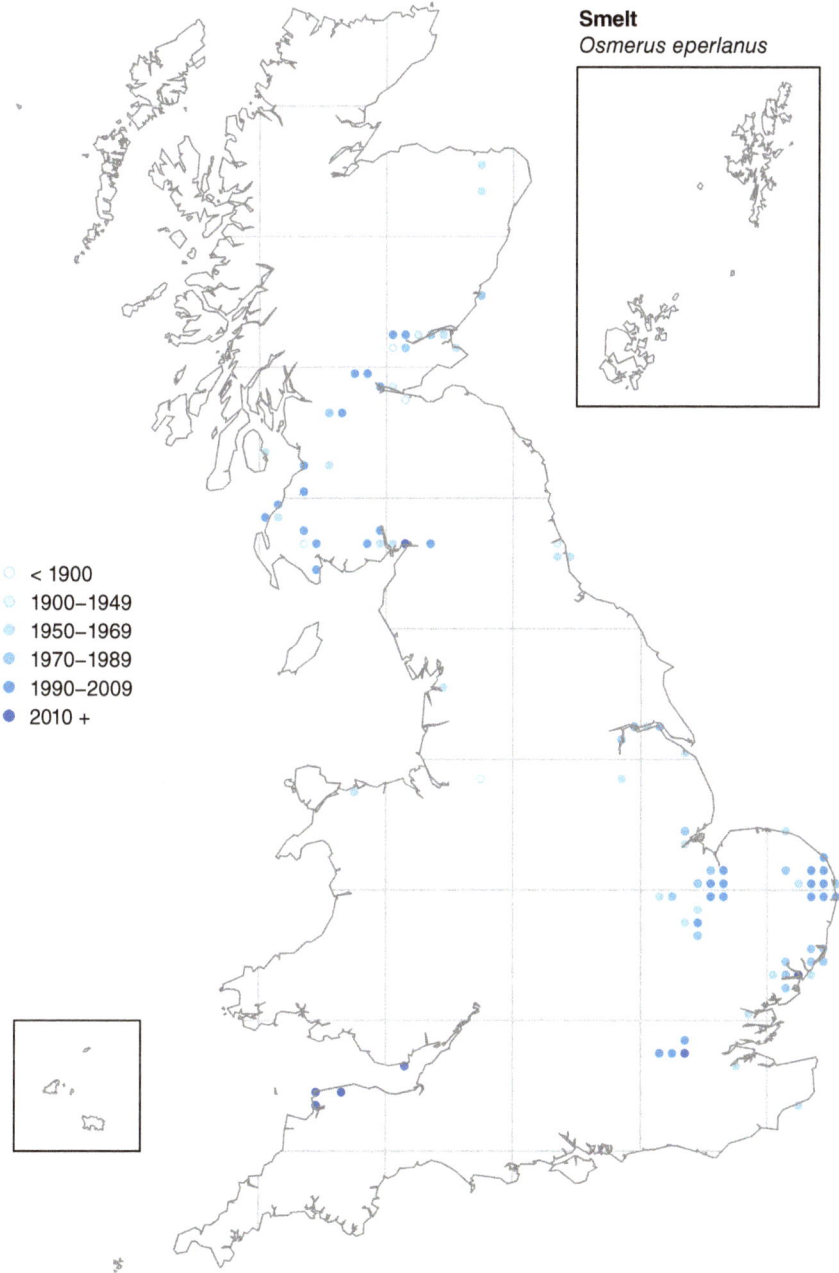

Smelt
Osmerus eperlanus

< 1900
1900–1949
1950–1969
1970–1989
1990–2009
2010 +

Fig. A1.11. Reported distribution of smelt across Britain (excluding Ireland).

Sand-smelt Distribution

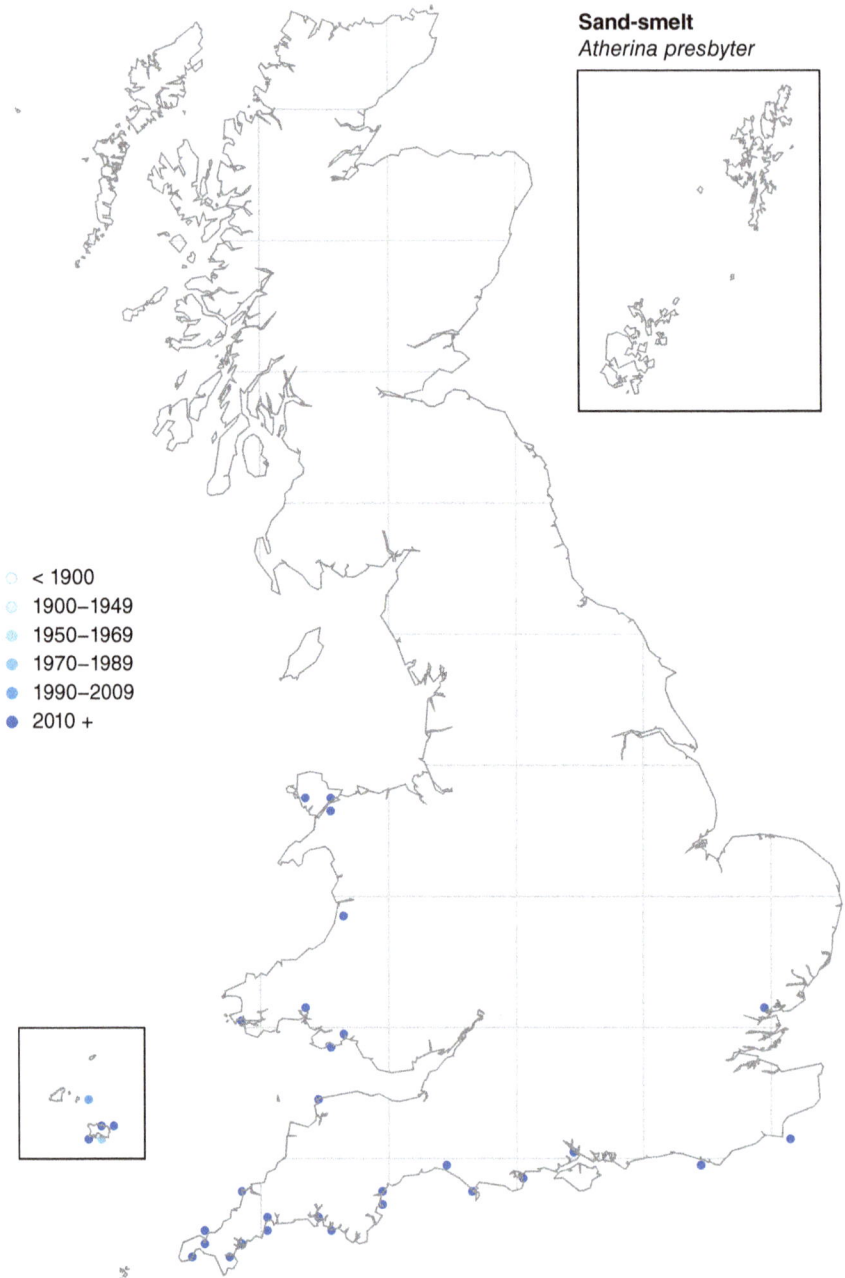

Sand-smelt
Atherina presbyter

< 1900
1900–1949
1950–1969
1970–1989
1990–2009
2010 +

Fig. A1.12. Reported distribution of sand-smelt across Britain (excluding Ireland).

Topmouth Gudgeon Distribution

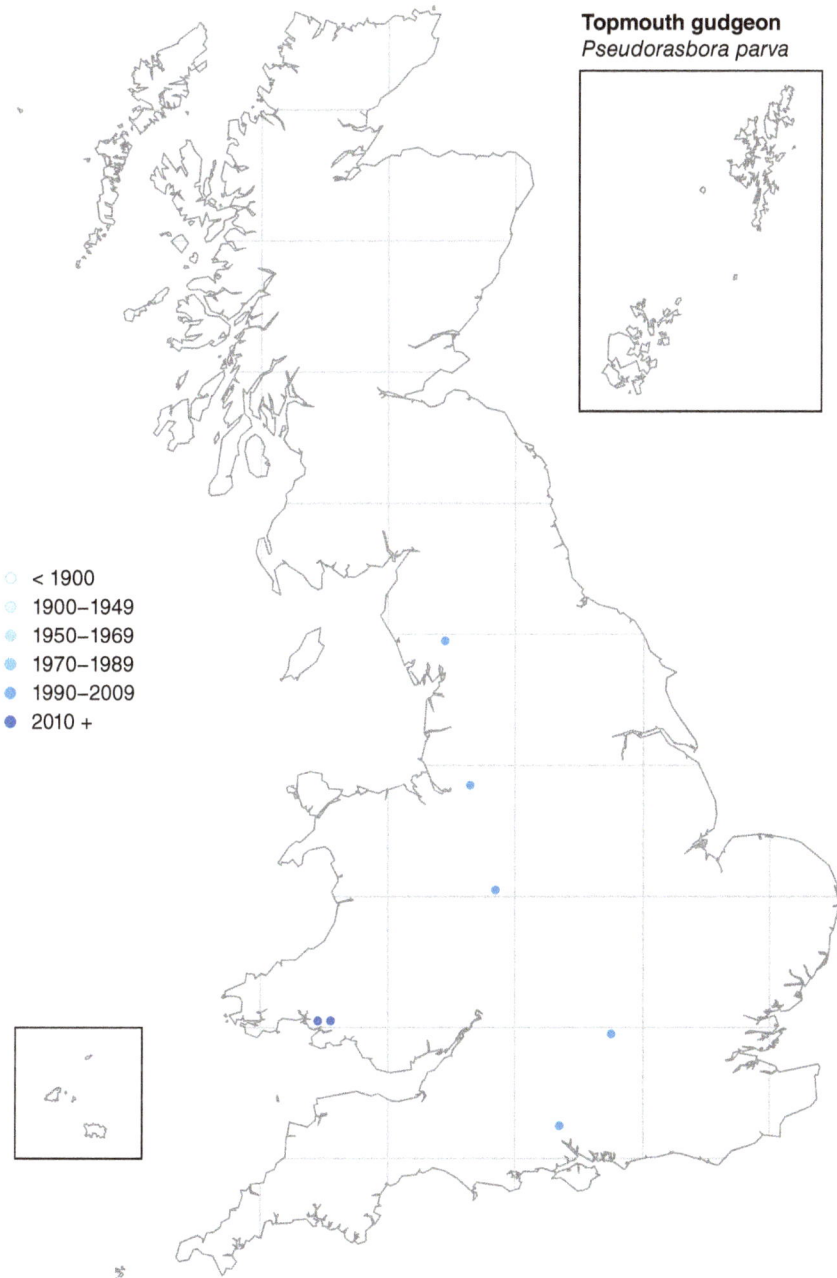

Fig. A1.13. Reported distribution of topmouth gudgeon across Britain (excluding Ireland).

Sunbleak Distribution

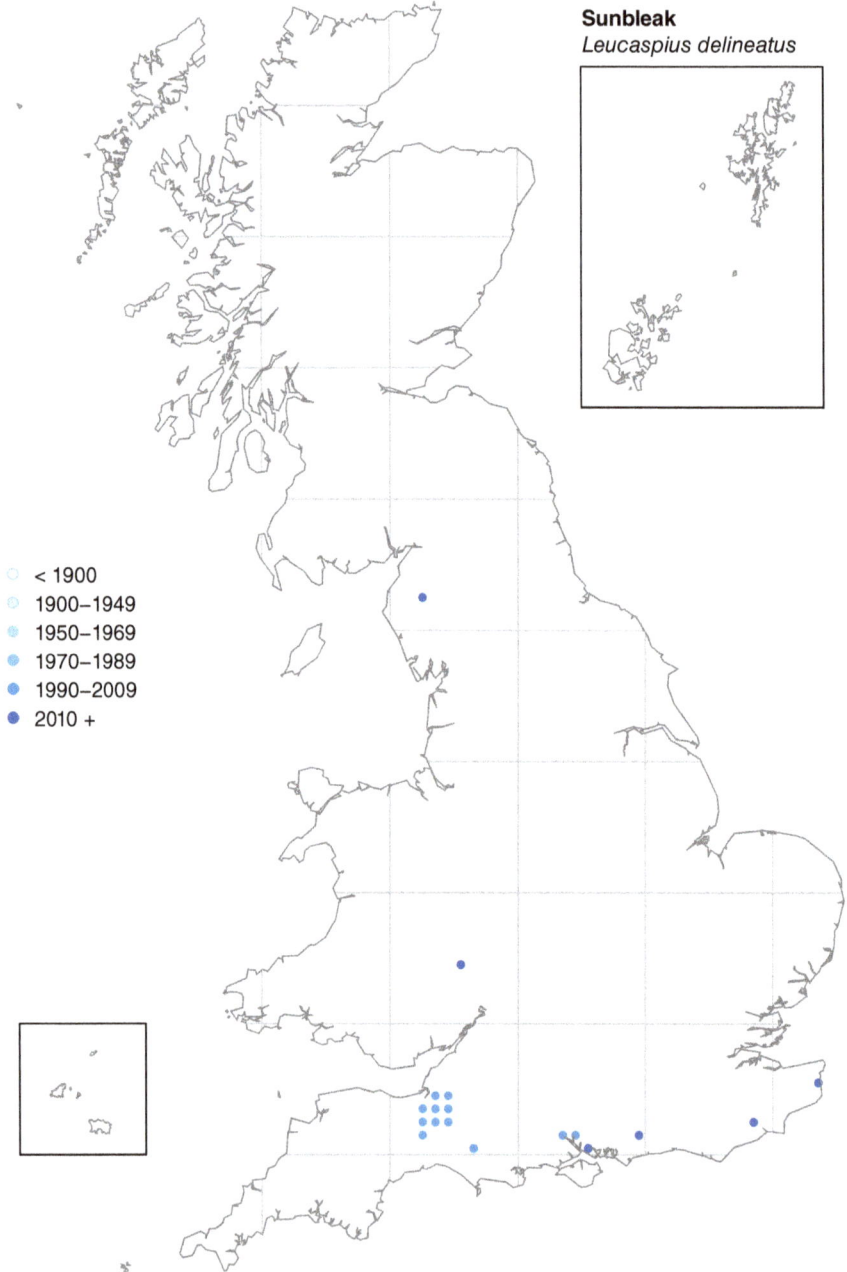

Sunbleak
Leucaspius delineatus

○ < 1900
◉ 1900–1949
● 1950–1969
● 1970–1989
● 1990–2009
● 2010 +

Fig. A1.14. Reported distribution of sunbleak across Britain (excluding Ireland).

Pumpkinseed Distribution

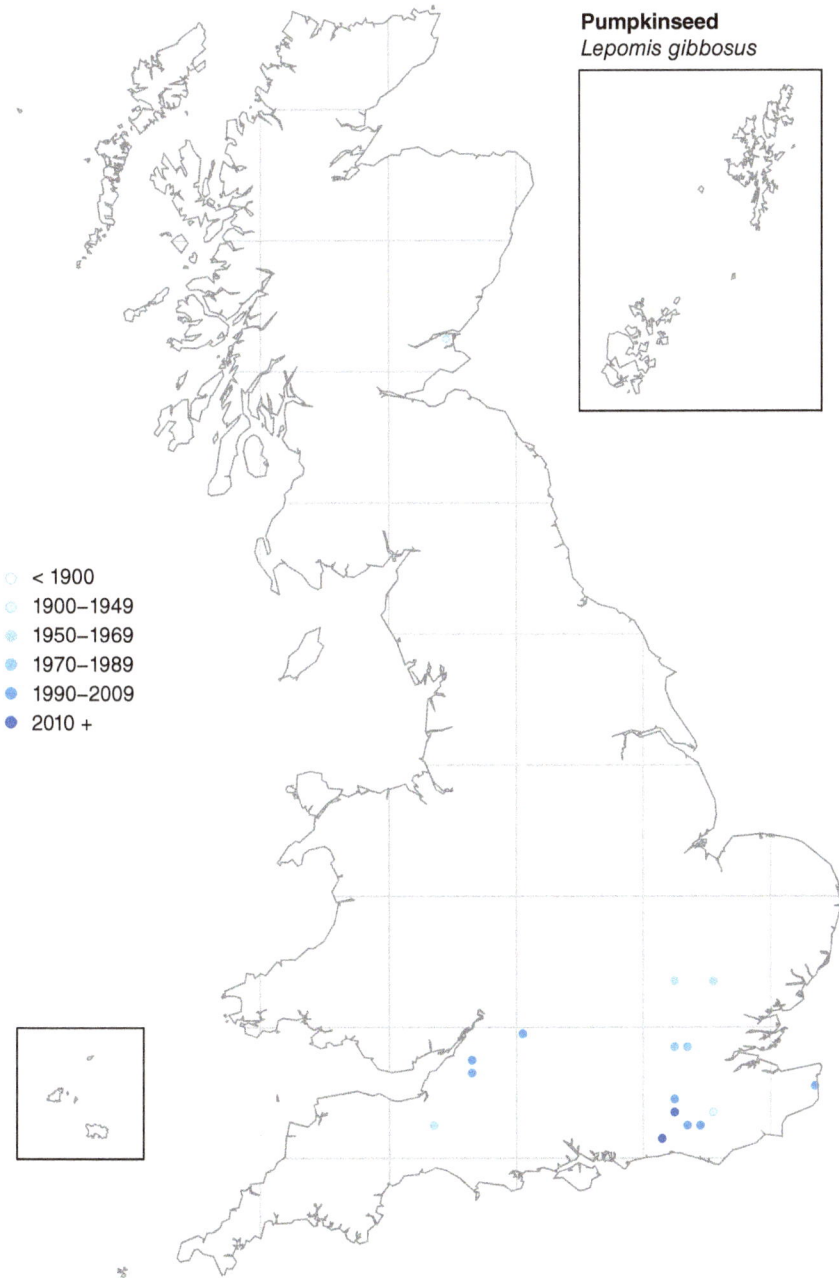

Pumpkinseed
Lepomis gibbosus

○ < 1900
◔ 1900–1949
◉ 1950–1969
◍ 1970–1989
● 1990–2009
● 2010 +

Fig. A1.15. Reported distribution of pumpkinseed across Britain (excluding Ireland).

Bitterling Distribution

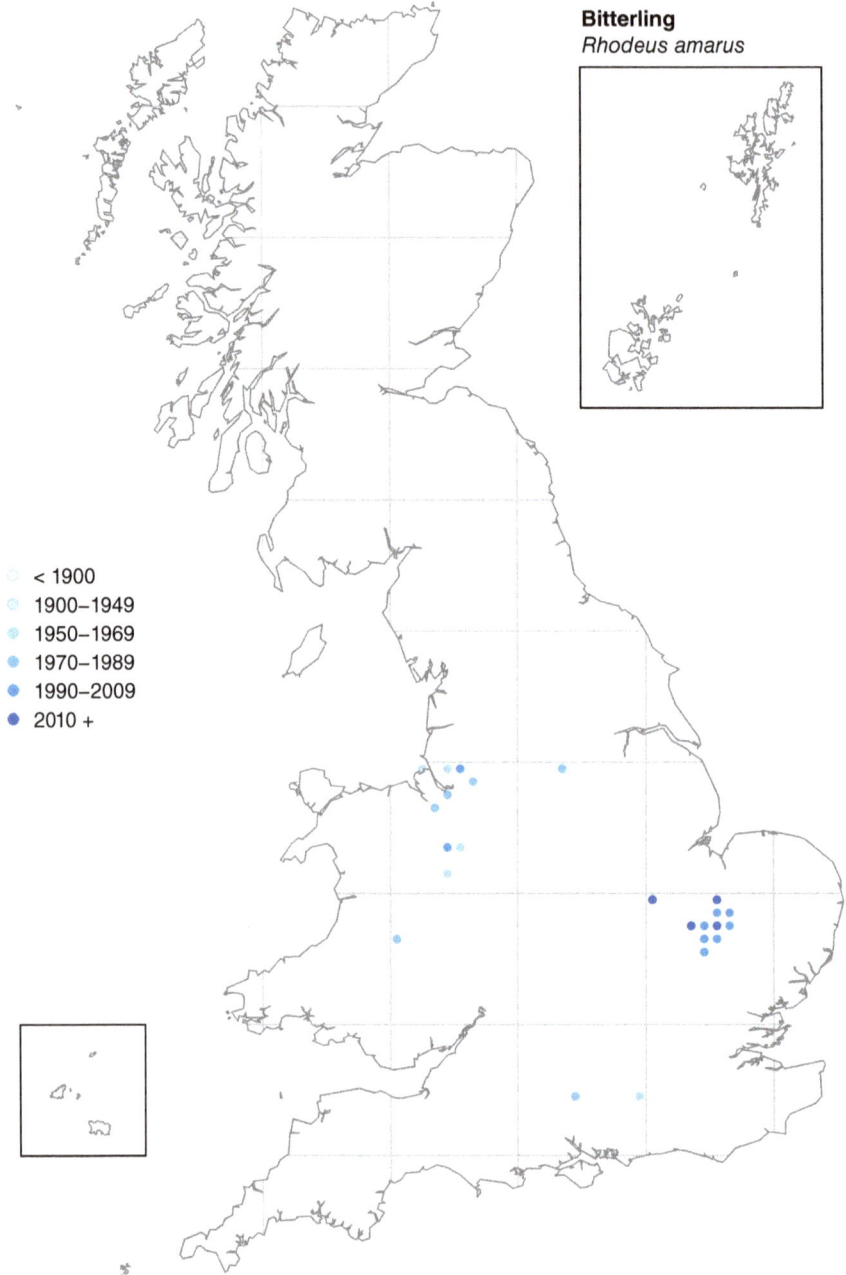

Bitterling
Rhodeus amarus

< 1900
1900–1949
1950–1969
1970–1989
1990–2009
2010 +

Fig. A1.16. Reported distribution of bitterling across Britain (excluding Ireland).

Goldfish Distribution

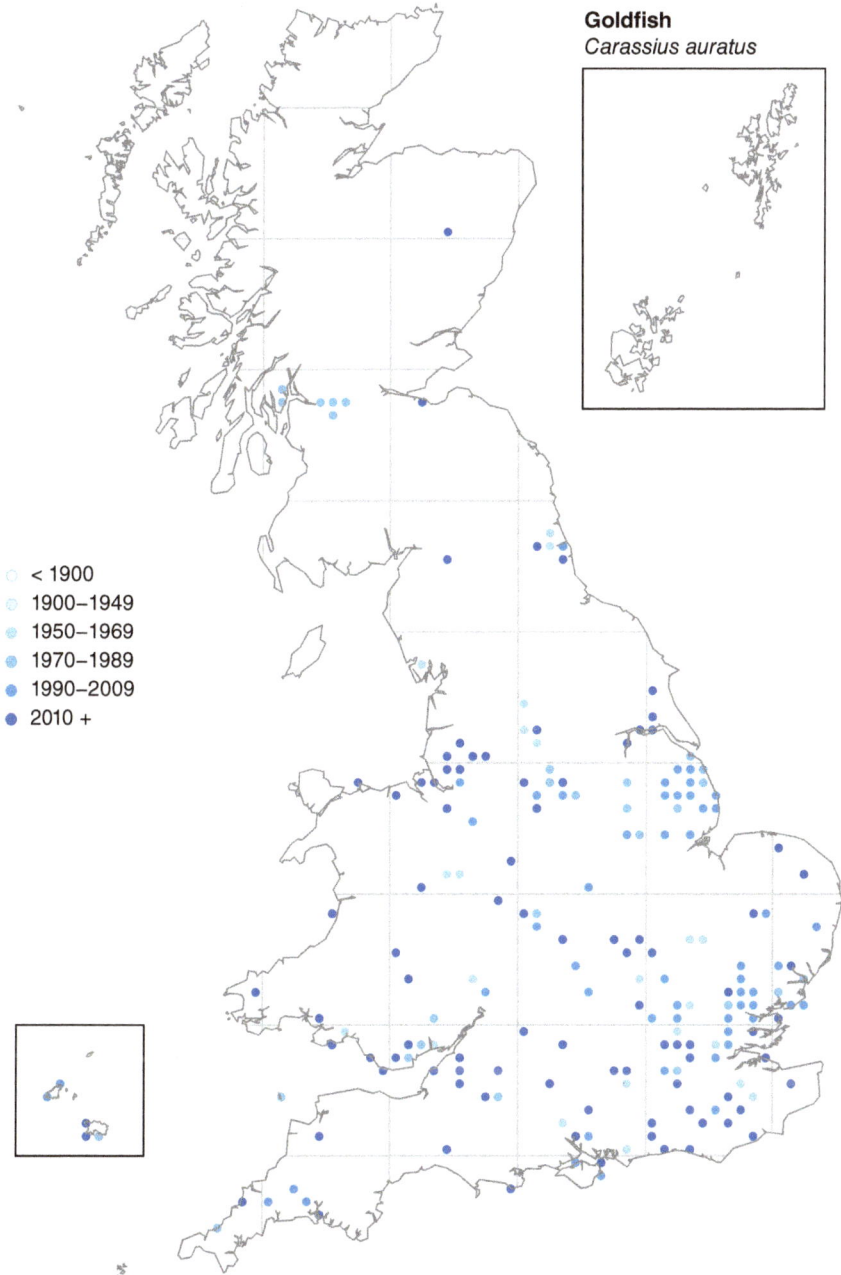

Goldfish
Carassius auratus

○ < 1900
◉ 1900–1949
◉ 1950–1969
◉ 1970–1989
● 1990–2009
● 2010 +

Fig. A1.17. Reported distribution of goldfish across Britain (excluding Ireland).

Gibel Distribution

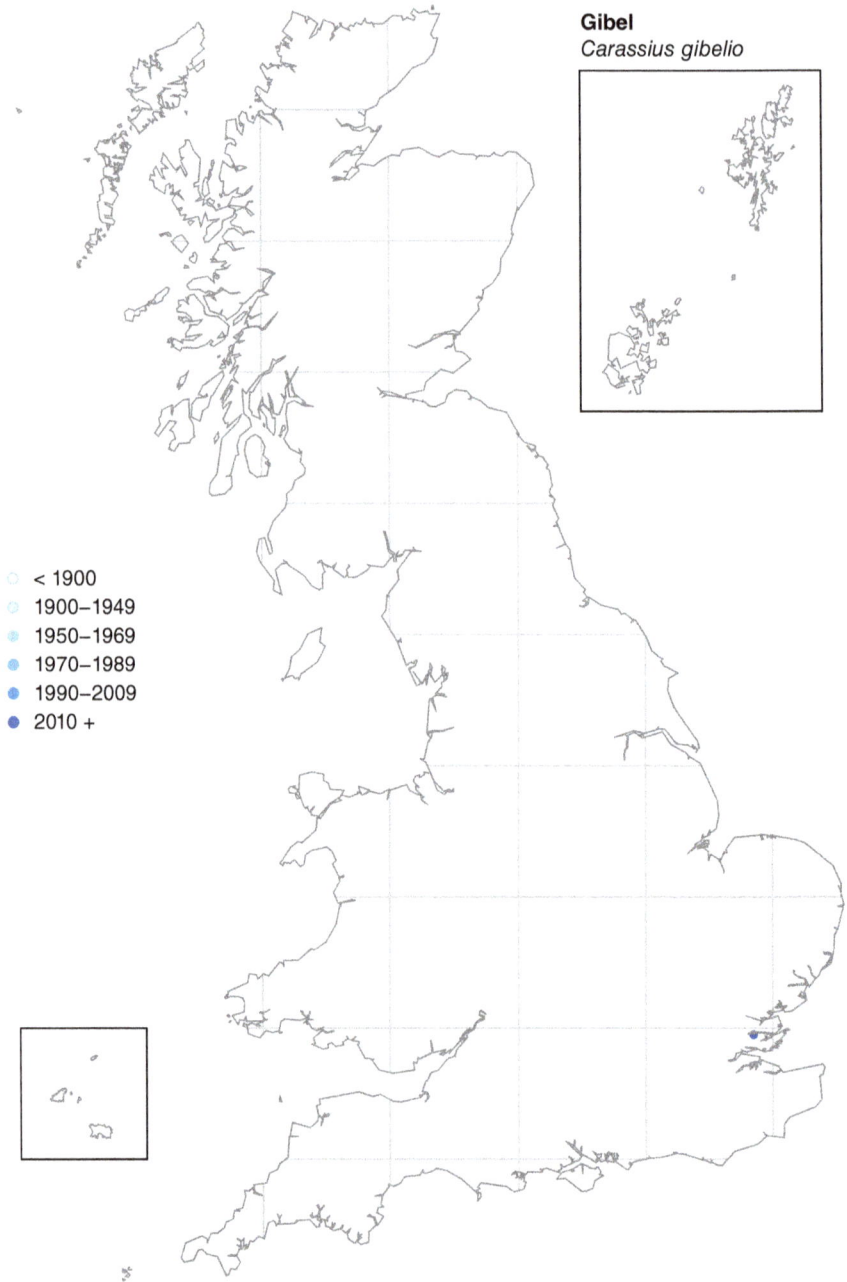

Gibel
Carassius gibelio

○ < 1900
◔ 1900–1949
◑ 1950–1969
◕ 1970–1989
● 1990–2009
● 2010 +

Fig. A1.18. Reported distribution of gibel across Britain (excluding Ireland).

White Sucker Distribution

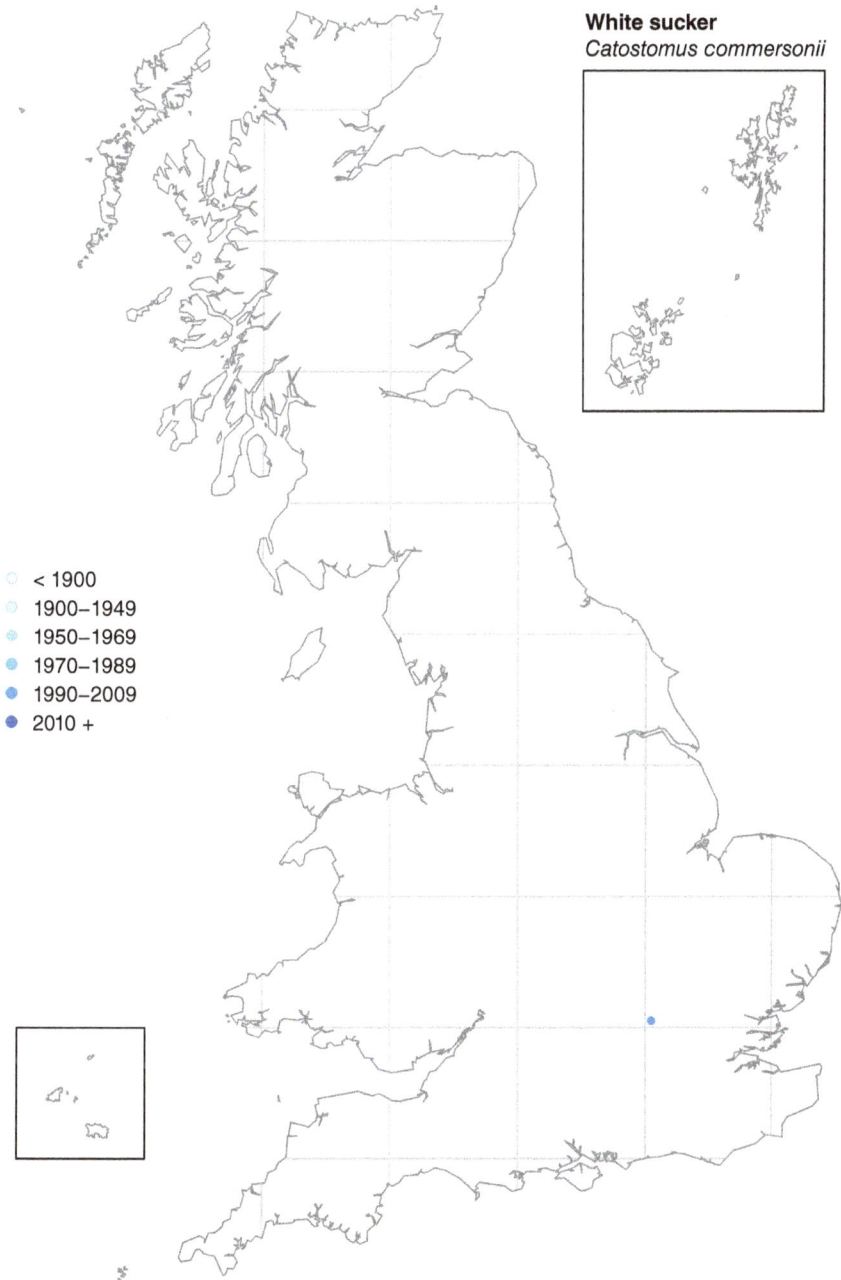

White sucker
Catostomus commersonii

○ < 1900
◔ 1900–1949
◑ 1950–1969
◕ 1970–1989
● 1990–2009
● 2010 +

Fig. A1.19. Reported distribution of white sucker across Britain (excluding Ireland).

Black Bullhead Distribution

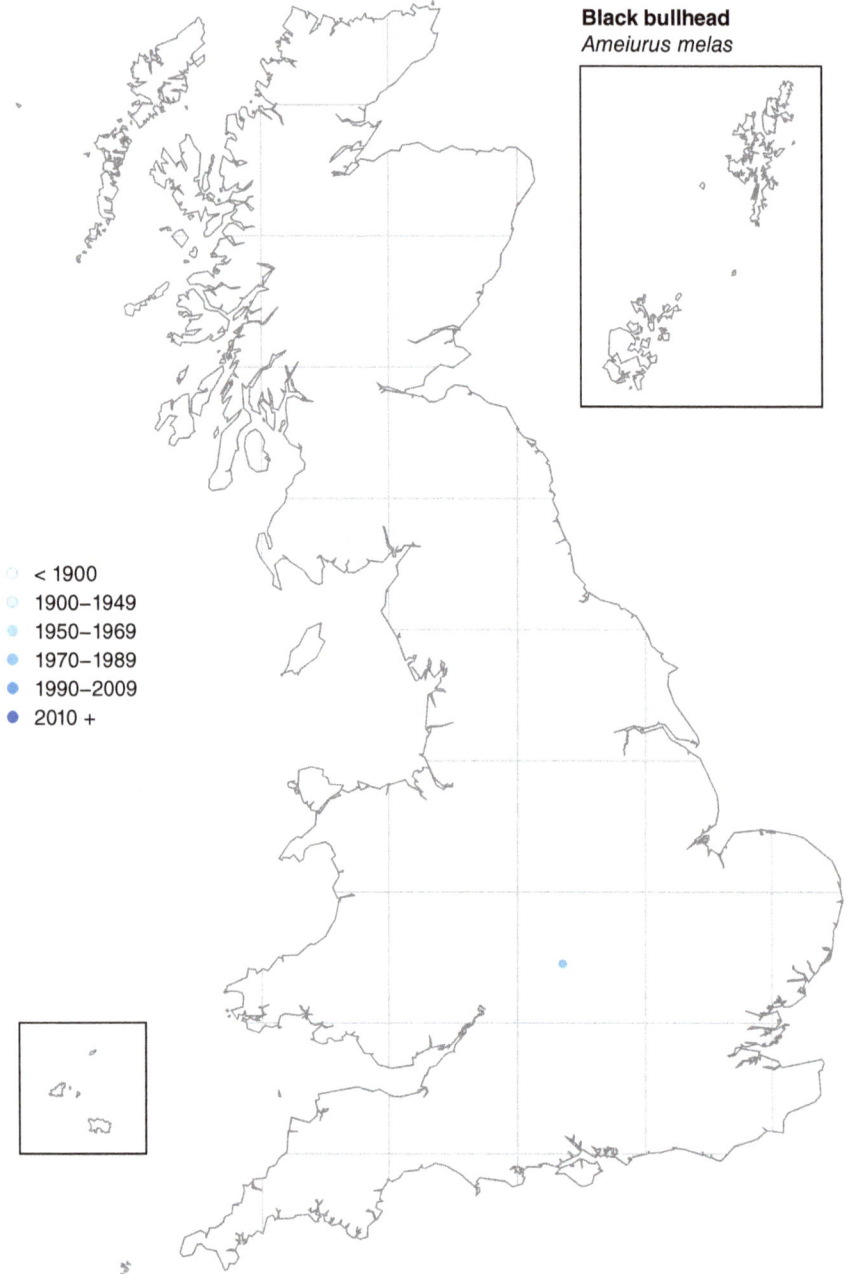

Black bullhead
Ameiurus melas

○ < 1900
○ 1900–1949
◉ 1950–1969
● 1970–1989
● 1990–2009
● 2010 +

Fig. A1.20. Reported distribution of black bullhead across Britain (excluding Ireland).

Index

Note: page numbers in italics denote figures.